Scaling Impact

Kusisami Hornberger

Scaling Impact

Finance and Investment for a Better World

Kusisami Hornberger
Dalberg Global Development Advisors
Washington, DC, USA

ISBN 978-3-031-22613-7 ISBN 978-3-031-22614-4 (eBook)
https://doi.org/10.1007/978-3-031-22614-4

Cover credit designers: Adriana Crespo & Paola Navarrete

This Palgrave Macmillan imprint is published by the registered company Springer Nature Switzerland AG
The registered company address is: Gewerbestrasse 11, 6330 Cham, Switzerland

About the Author

Kusisami Hornberger Kusi is a Partner in Dalberg Global Development Advisors, based in the Washington, DC office. In addition to serving as Dalberg's Global Knowledge Lead, Kusi also co-leads Dalberg's Finance & Investment Practice and is particularly passionate about the use of innovative finance and technology to accelerate achievement of the United Nations Sustainable Development Goals (SDGs). Over the course of his career, Kusi has led more than 200 advisory projects on a wide range of topics related to blended finance, financial inclusion, impact investing, impact measurement and management, and innovative finance with leading bilateral and multilateral donors, development finance institutions (DFIs), family offices, foundations, and private investors around the globe.

Prior to joining Dalberg, Kusi was Vice-President of Investment Research at Global Partnerships, an impact-first investor with investments across Central/South America and East Africa. He also has experience working as a management consultant at Bain & Company in South America, as an Investment Officer with the International Finance Corporation, and as consultant with TechnoServe Inc. in East Africa. He started his career as a Peace Corps Volunteer in Tanzania, where he served two years as a math and computer science teacher in the foothills of Mount Kilimanjaro.

Kusi holds a Master of Business Administration from INSEAD Business School in Singapore, a Master of Public Administration in International Development from the Harvard Kennedy School, and a Bachelor of Arts in Economics and International Relations from the University of Pennsylvania. Kusi was born in Lima, Peru, grew up in West Philadelphia and currently resides in Bethesda, Maryland, with his wife and daughter.

Acknowledgements

Writing a book has taught me that no book is really the product of any single individual's endeavors or efforts. This book is the product of several direct contributors and many more supporters for whom I am deeply grateful. I would like to start by thanking Fernando Martins, who through our various one-on-one chats encouraged me to "just get started," which led me to begin the journey, as well as Steve Feldstein, who encouraged me to seek a publisher. I'd also like to thank Tula Weis and the entire Palgrave Macmillan team, who helped me formalize my dream and gave me a pathway to seeing this book come to life.

Once I got started, I would never have finished had it not been for a terrific trio of support I had from start to finish from a research assistant, copyeditor, and designer. First, I'd like to thank Catherine Ross, who agreed to be my research assistant, and whose constant companionship and good advice as I pieced the book together over several months was invaluable. Next, I'd like to thank Susan Boulanger, who was my copyeditor in chief and reviewed every single chapter and citation in the book and who has been there as my editor for many more pieces of work than probably she would have liked. Finally, I'd like to thank Adriana Crespo, who served as the designer for the book cover and the many figures, tables, and graphics that appear throughout the book. Thank you for your creativity and beautiful outputs, Adriana. I am so deeply appreciative.

I also thank the six esteemed guest authors: Bjoern Struewer, Chris Jurgens, Joan Larrea, Nicholas Colloff, Olivia Prentice, and Sasha Dichter. All of them enthusiastically agreed early on to participate, even when the book was just an idea. They not only happily volunteered and provided timely contributions, they also helped to push my thinking and without doubt made the book much better than it would have been without them. Thank you so much for agreeing to join me on this journey and for the expertise and insights you have graciously shared here.

I'd also like to thank the many peer reviewers I solicited throughout the process to look at different sections of the book, notably Abhilash Mudliar, Aunnie Patton Power, Belissa Rojas, Dan Waldron, Eelco Benick, Mathieu Pegon, Mike McCreless and Tara Murphy Forde. Your generous contributions of knowledge and experience strengthened my content and writing immeasurably, and I am so grateful that you took the time to help me.

In addition, I'd like to thank Dalberg for being not only a terrific place to work but for supporting me as I took on this project, and the many current and former colleagues who shared thoughts, reviewed, and added color to help make the book stronger. Notably I'd like to thank Ben Stephan, Bianca Samson, Camila De Ferrari, Eric Grunberger, Erin Barringer, Georgina Muri, Greg Snyders, Fabiola Salman, Flavia Howard, Jeff Berger, Ines Charro, Kaylyn Koberna, Kristina Kelhofer, Laura Amaya, Laura Herman, Lucie Gareton, Mackenzie Welch, Marcos Paya, Mark Pedersen, Mette Halborg Thorngaard, Nick Whalley, Shyam Sundaram, Sophie Gardiner, Rachna Saxena, Stig Tackmann and Yana Kakar.

A lot of credit goes also to the many wonderful clients I have worked with through the years. Many of the advisory projects I had working with you served as direct input or inspiration for the content that you find in this book. Specifically, I would like to thank Adriana Suarez, Alexander Dixon, Amit Varma, Amy Lin, Anne Katharine Wales, Beth Roberts, Brian Milder, Caroline Bressan, Chris McCahan, Clara Colina, David Milestone, Deidra Fair James, Drew Von Glahn, Elizabeth Boggs Davidsen, Harry Devonshire, Hillary Miller, Irene Arias Hofman, James Hallmark, Jamie Anderson, Julia Kho, Karen Larson, Lorenzo Bernasconi, Luca Etter, Meghan Curran, Mikael Hook, Paurvi Bhatt, Peter Tropper, Priya Sharma, Richard Ambrose, Rob Schneider, Robert Haynie, Robin Young, Sharon D'onofrio, Songbae Lee, Stephanie Emond, Susie Qian, Tim Hanstad, Tim Rann, Urmi Sengupta and Yolanda Banks. Thank you for giving me the opportunity.

Also, my thanks are due to the many supporters who reviewed final versions and provided their perspectives and praise, if merited—most notably, Aunnie Patton Power, Brian Requarth, Brian Trelstad, Catherine Clark, Claudia Zeisberger, Gernot Wagner, Jacqueline Novogratz, Neil Gregory, and Santitarn Sathirithai.

I want to especially thank Brian Requarth, Aunnie Patton Power, Marco Dondi, and Morgan Simon, whose own publications served as inspiration for this book, as well as the many friends who inspired the examples used here, notably Carlos Velasco, Gabriel Migowski and Xavier Saginieres. As well as the many friends without I have engaged in thoughtful debate through book reading clubs and inspired many of the ideas in this book including Alex Evangelides, Ami Dalal, Caroline Bressan, Chris Walker, Lauren Cochran, and Songbae Lee among others. A special thanks to the many friends who responded to numerous requests for feedback on this book project, including Alessandro Zampieri, Andrew Siegel, Asaph Glosser, Daryl Weber, David Schlosberg, Felipe Valencia, Humberto Laudares, Jolene Yap, Knud Lueth, Mark Quandt, Mila Lukic, Matt Asada, Lukas Steinmann, Oliver Gillin and Steve Feldstein among others.

Thanks also to the numerous mentors and teachers who shaped who I am and the thinking that went into this book, from my early Quaker education to my formative early years in university and on the job: Andre Leme, Armando Heilbron, Bruce McNamer, Cecilia Sager, Charlotte Davis, Cheng Davis, Herb Kerns, Claudia Zeisberger, Damien Shiels, Dani Rodrik, Erica Field, Fernando Martins, Henrik Skovby, Irene Arias Hofman, Jere Behrman, Jeremy Siegel, Josephy Stiglitz, Larry Westphal, Lawrence Klein, Mark Coffey, Patrick Turner, Peter Bladdin, Pierre Hillion, Rayond Hopkins, Ricardo Hausmann, Rick Beckett, Robert Bates, Rodolfo Speilmann, Stephen Golub, Stephen Moshi, Veronica Chau and Yana Kakar.

Finally, I'd like to thank my family, most importantly my wife, Laura, and daughter, Ariana, for their patience in allowing me to work on this crazy project for many a weekend and free moment that I really should have been spending with you. And to my parents, Steve and Nancy, and my sister, Ch'uya: thank you for your numerous responses to requests for feedback and for a lifetime of your support. This book is a dream of mine, but it is also a gift to you. Thank you.

Preface

As I sit down to write this preface, the world is beginning to emerge from the global coronavirus pandemic. The last three years have taught us the importance not only of being better prepared for the future pandemics that are sure to come but also of making sure we understand, communicate, and take advantage of the latest innovations from science and technology that have allowed us to return to our lives. That is not to say that technological innovations like new vaccines are a silver bullet for achieving a better world—they don't ensure equity or the distributional effects of their advancement—but certainly if we take advantage of the good parts of such innovations and protect against their risks, I believe we can collectively use them to make substantial positive progress in the world. Technological innovations can help us not only to better avoid and protect ourselves against future pandemics, but also to avoid and reduce the risks of climate change and even to solve poverty and end systemic inequality.

As a graduate student, I spent a lot of time thinking and reading about and trying to understand what leads to economic growth. One of the most important things I learned about was the concept "total factor productivity," or TFP. The concept was developed by Robert Solow, a Nobel prize–winning economist and professor emeritus at MIT.[1] As he modeled the growth of firms, he found that not all output growth at the national or industry level can be explained by the accumulation of traditional inputs such as labor and capital. Even after accounting for these, a residual factor behind growth remained unexplained. That residual factor, now known as TFP, is in essence productivity or technological innovation gains that allow growth in output beyond invested input.

Later on, after a series of professional experiences, I saw that not all investments were created equal and that some did more good than others. Others, unfortunately, even caused unintended harm to some stakeholders or to the environment in ways that were not understood before the transactions took place. Combined with more and more reading about the need to shape investments in growth so that they benefit all stakeholders, this understanding made me realize the limits to seeking growth alone. The truth is, in practice, not all innovations that create growth are good. Investments in economic growth and innovation must be sustainable. Following Kate Raworth's theory of doughnut economics, we must also think about growth within our planetary and social boundaries.[2] We must create and invest in a world where the needs of all people are met without overshooting the earth's ecological ceiling or using up all its resources.

That is what this book is about. It is about innovations—not in science or medicine or national output, but innovations in the practice of finance and investment that can create sustainable outcomes. It is about how we can evolve our thinking about the rules for using capital and capitalism as forces for good and solving our world's most pressing challenges: not by simply changing the amount of capital we put in, or the effort (labor) we supply, but by rethinking and innovating on some of the central tenets of capital. It is about how investors who place capital for return can truly prioritize creating value for all stakeholders, placing that goal at the center of their investments. It is about how governments and philanthropists can best use their resources to catalyze innovation. It is also about how finance itself can be designed to consider its users' and not just its providers' needs.

While encountering the terms "Solow growth model and TFP" and "Raworth's doughnut economics" might scare you at first, this is not a technical book. I wrote it to help demystify some new finance and investment topics, such as impact investing and blended finance, for those less familiar with the concepts. It provides perspectives on how to use those tools successfully, offering insights from and for practitioners. It is written in plain, easily accessible language for students and learners, and it is filled with insights and examples that will be relevant to more advanced finance and investment professionals as well. It is also a call to action for anyone who cares about solving global problems, believes in our capitalist system, and wants to use it to solve its problems—like I do.

So why write a book now? The fact is I love to research, write, and share knowledge. While I never became a professor, one of the things I still love best is coming up with unique insights and sharing them. I also discovered that over my 20-plus years as a professional working on economic development, development finance, and impact investing, as both an investor and advisor, I had started to accumulate knowledge that was useful to others and interesting insights that resonated with them.

> *To solve the world's hardest problems, spend twice the time communicating the solution than you did coming up with the technically correct answer.*
> —Kemal Dervis, visiting lecturer, John F. Kennedy School of Government, Harvard University, 2006[3]

I heard Kemal Dervis speak these words while I was in graduate school, and remembering them later, I started to write short articles and blogs about what I was learning from my reading, thoughts, and interactions with others sharing my interests and concerns. I put them out in the world with the goal of striking a response among colleagues and others. It started with my blog and website, "Social Impact Business Models," on which I highlighted different profitable business models, discussed how they made money, and outlined how they contributed solutions to a societal problem. Later I began to write pieces directly on LinkedIn, synthesizing some of my professional and personal research on impact investing. The reaction, particularly to the latter, was enthusiastic. I started to get thousands of views a day and often hundreds of likes and shares of some of my posts. This convinced me I had finally found something unique and useful to share that might even deserve a wider audience.

I decided to take advantage of the pandemic and our long working-from-home situation to collect all those perspectives and piece them together into a single book or resource that could be accessible to anyone. Deciding that it would be even more helpful if I combined my writings and perspectives with those of other leading practitioners in impact investing, results-based finance, and blended finance, I invited some of the field's leading intellectual lights and practitioners whom I had long admired to join me in the areas of the book particularly relevant to their contributions to innovations and advancements in finance and investment for good. These contributors include Sasha Dichter, Chris Jurgens, Joan Larrea, Olivia Prentice, Bjoern Struewer, and Nicholas Colloff. I believe this book and its insights about strategies, structures, and practices in blended finance and impact investing are needed now more than ever, given the unprecedented challenges we currently face. I hope you will find the book useful in bringing needed shifts toward impact in practice of finance and investment to scale.

Brownsville, VT, USA
February 2022

Kusisami Hornberger

Notes

1. Solow R (1956) A Contribution to the Theory of Economic Growth. *The Quarterly Journal of Economics* 70(1): 65-94.
2. Raworth K (2017) *Doughnut Economics: Seven Ways to Think Like a 21st-Century Economist*. Chelsea Green Publishing, White River Junction.
3. Guest Lecture in Michael Walton's course on Policy Design and Delivery at Harvard Kennedy School.

Contents

List of Figures

1

Introduction

The global challenges we confront daily can feel overwhelming

Some mornings it is hard not to feel overwhelmed. Global challenges of poverty, climate change, and inequality affect us daily. The size and severity of these and other global challenges can feel daunting whenever we think about them.

When I awaken each morning and make a cup of coffee, for example, I am reminded that half of the world's smallholder farmers—producers of the coffee beans I am grinding and many other products important to us—live in poverty.[1] Hundreds of millions of rural families continue to live off the production and sale of unprocessed cash crops, like coffee grown on small-scale farms. These families often have between two and five children, live in semi-formal structures, and survive season to season on the meager earnings they cobble together from farming and any non-farm income they can generate. While I feel grateful that I can access global supply chains that bring coffee from those farms to my table, I also feel saddled with guilt that we haven't figured out how to give those farm families access to better and more stable incomes.

When I open my phone and scan the news and my social media feeds, I often encounter news of an ever-increasing number of natural disasters and extreme weather events in my own country that are caused by rising temperatures. Whether droughts, heatwaves, or wildfires in western United States,

K. Hornberger, *Scaling Impact*,
https://doi.org/10.1007/978-3-031-22614-4_1

deadly tornadoes in the Midwest, or the increasingly negative impact of the hurricanes along the East Coast and Gulf of Mexico, the deluge of news about potential disasters is year-round. According to the *New York Times,* in 2021 temperatures across the United States set more all-time heat and cold records than in any year since 1994.[2]

The same pattern is seen around the globe. The rise in the number of severe floods and storms has been startling. According to the United Nations, over the last 20 years, as compared to the previous 20, there has been a staggering rise of nearly 75 percent more extreme weather events.[3] The same study showed that those major natural disasters killed 1.23 million people and resulted in $2.97 trillion in global economic losses. Worse, an analysis of 357 peer-reviewed studies found that 70 percent of 405 extreme weather events around the globe were made more likely or severe by human-caused climate change.[4] Data like these often make me—and many like me—feel powerless: The actions required to reverse the effects of climate change seem too great and far beyond my control.

Finally, heading out the door to drop off my daughter at school, I am reminded how fortunate we are that she can attend in-person school: So many children worldwide—particularly those in disadvantaged communities—have seen their learning disrupted by the pandemic. COVID-19 was a wake-up call showing people everywhere how the negative shocks of crisis are disproportionately experienced by those least able to deal with them. Data have now been gathered revealing that learning losses were greater for students of lower socioeconomic status, including in countries from Ghana to Mexico to Pakistan[5] and minority populations in the United States.[6]

This is just a small sample of the daily reminders many of us encounter that can leave us with a sense of hopelessness and of our failure to do enough to help make the world a better, more just place. The United Nations estimates that it will cost between five and seven trillion dollars per annum to achieve the Sustainable Development Goals (SDGs) by 2030.[7] Currently, driven primarily by government spending, we finance less than half of this cost. Much more investment—roughly the size of Germany's gross domestic product (GDP), or almost four trillion dollars—is needed annually. The hope is that private players can help close the gap and contribute to global efforts to achieve the SDGs.

Is capitalism failing us? Does it need to be reimagined?

Just as we begin to believe that we must rely on private actors and private capital to be our saviors, we learn it is not so simple. Misuse of capitalism's tools has too often undermined rather than advanced progress on social issues. Private individuals in positions of power lead the agendas, and they can bend the rules in their favor.

Just as we were starting to believe that the impact investing movement would "mainstream" through billion-dollar private impact investing funds such as the TPG Rise Fund, for example, bad actors were found. The managing director of that fund, Bill McGlashan, along with several other wealthy elites, was caught and convicted[8] for paying bribes to fix his son's exam results and gain his admission to the University of Southern California. In essence, he subverted the system for his family's personal gain and stole an admission spot at a good college from another deserving student who may not have had the same access to resources and advantages.

Similarly, but on a global scale, another private equity impact investor has drawn the ire of many for his actions, subverting the impact investing narrative for his personal gain. Arif Naqvi, the founder and chief executive of The Abraaj Group, was once heralded as having proven that commercially driven private equity funds could be successfully invested in emerging markets with a double or triple bottom line. Investing in healthcare, water, and basic infrastructure in places like Pakistan, Senegal, and Turkey, Naqvi built a billion-dollar empire of funds with resources from tier capital sources, including leading development finance institutions, pension funds, and even the Bill and Melinda Gates Foundation. Yet, when it was discovered that Naqvi was using his position of power to siphon funds illegally from one place to another to feed his personal spending needs, the "house of cards" eventually fell apart, as detailed in the book *The Key Man.*[9]

These scandals were a siren call to many about the true intentions of leadership at "impact investing" funds like TPG Rise and The Abraaj Group. If wealthy elites like Bill McGlashan and Arif Naqvi are willing and able to so blatantly cheat the system for personal advantage, why wouldn't they or others in a similar position cheat and/or manipulate impact investments in the private markets to their advantage as well? It appears Mr. McGlashan and Mr. Naqvi were driven more by ego and by the narrative of achieving social and environmental goals than by achieving the impact goals themselves. This isn't changing the status quo, and it certainly doesn't help build the trust needed to scale private capital solutions to meet global challenges.

It also suggests that "capture" is a problem in the private markets, rather than just for governments, as exposed in public choice theory—a theory for which James Buchanan won a Nobel prize in 1986,[10] which explains government decision-making can result from actions by individual, self-interested policymakers. It should give us all pause about the reliability of unchecked market-based systems to deliver the results we desire. We need to realize that the rules that shape markets are not "naturally occurring" nor do markets in themselves give rise to fair competition, as James Buchanan leads us to believe. Rather, markets and the actors within them, particularly in private markets in low- and middle-income countries, are designed and manipulated by people and powerful interests in ways that often worsen economic, social, and environmental outcomes.[11]

As a case in point, in the United States—an economy that many still believe is the most competitive and free of all major economies in the world, and where you can realize the "American dream"—intergenerational economic mobility has become stifled. For many, it's not hard work or individual achievement, but rather race and the neighborhood you grow up in that have much more to do with the likelihood of jumping from the bottom quintile of incomes to the upper quintile.[12] Even when controlling for all other factors, such as effort, type of school you attended, or aptitude, you are still ***less*** to see your lot in life improve likely in the United States than you might be in countries with higher taxes and larger governments, such as many in Europe.

Worse, there is also evidence that entrepreneurship rates have slowed, and the amount of wealth concentrated among the "winners" in the economy has reached historical levels.[13] Many private sector economic titans are incentivized to spend to increase productivity rather than to pay or retain workers. At the same time, most of the people whose work generated those wealth gains are left behind, especially in communities of color. The result has been an explosion in economic insecurity across the United States and in much of the industrialized world. Middle-class skilled labor has evaporated, and income inequality and the racial wealth gap have destabilized many economies, undermining our collective sense that we live in a fair and just society.

Yet real and measurable progress is being made

Despite the fears and the sometimes-overwhelming conviction that things couldn't be worse, evidence actually indicates that the contrary is true. Humankind is making positive progress against challenges that once felt insuperable, and there truly has never been a better time than today to be alive on planet Earth.

Consider poverty—the issue I have most closely focused on in my career. We are making significant and rapid progress to end poverty. The projected 2022 rate of extreme poverty (people living on less than $2.15 a day) is less than one-third of what it was 20 years ago (8.5 percent versus 26.9 percent).[14,15] In fact, over the last 25 years, more than a billion people have lifted themselves out of extreme poverty, and the global poverty rate is now lower than it has ever been in recorded history.[16] This may be one of the greatest human achievements of our time. It is possible that if current trends continue and extend into Sub-Saharan Africa, poverty will be eradicated everywhere.

Some point out that poverty is only one part of the equation: If inequality is rising, eliminating absolute poverty may not be enough. Income inequality between countries has improved as well over the past several decades, however. Because average incomes in low- and middle-income countries are rising faster than in high-income countries, the average income gap between the richest 10 percent of countries and the poorest 50 percent dropped from 53 to 38 times over the past four decades.[17] And while within-country inequality has problematically increased in most countries over the past three decades, it has improved in some regions, such as Latin America and the Caribbean (albeit from a high level).

When looking at inequality of opportunity by gender or race, we also see some signs of progress. On the gender front, wage equality and the proportion of women among skilled professionals has increased in recent years.[18] In most OECD countries, there is little—if any—difference in unemployment rate between men and women.[19] Further, gender gaps in these countries in education attainment, health, and survival are nearly closed.[20] While certainly not enough, these are positive signs of progress.

On the racial inequity side as well, while much more remains to be done, here too we see some signs of progress. Take, for example, data showing that between 1986 and 2016 in the United States the proportion of Latinx families with zero or negative wealth dropped by nearly one-fifth, from 40 to 33 percent.[21] Similarly, between 1983 and 2016, the rate of homeownership among Latinx families increased by 40 percent.[22] Again, this is not nearly enough progress, but these are signs of progress on a systemic problem in the United States.

To take another example, consider global health. This is perhaps the area in which the most positive progress has been made in the last 50 years. Life expectancy is rising globally as child mortality has halved since 1990.[23] Progress in Sub-Saharan Africa, where maternal mortality has declined by 43 percent between 1990 and 2015.[24]

Before the COVID-19 pandemic[25] nutrition rates had also improved as world hunger has fallen over the past several decades. Worldwide, the percent of undernourished people declined from more than 12 percent in 2004 to less than 9 percent in 2018. The rate of stunting in children under age five had also declined by more than 4 percent between 2012 and 2020.[26]

Best of all, while new health risks continue to emerge, such as the coronavirus pandemic, we have succeeded in eliminating many diseases that previously killed millions: Smallpox, and guinea worm disease have been completely eradicated. The number of new HIV/Aids infections per million people has more than halved from 1997 to 2020,[27] and the incidence of malaria declined by 25 percent between 2000 and 2019.[28] We are also on the verge of having a malaria vaccine, which would stop this disease from ravaging the planet and particularly the poorest populations living near the equator.

Finally, if we look at climate change—the greatest collective threat we face—there is also some positive progress being made. Global warming projections are improving. Today's 2100 global warming projection is more optimistic than those made prior to the 2015 Paris climate agreement, at 2.7 degrees Celsius rather than 4 degrees Celsius, reflecting significant (although not enough) progress.[29] At the same time, strengthened resilience and adaptive capacity to climate-related disasters in some sectors and communities has meant that, although there were nearly five times more natural disasters in the last 20 years, three times fewer disaster-related deaths occurred over this time period, largely due to improved early warning systems and disaster management practices.[30]

We also see momentum and progress being made with renewable sources of energy. Between 2010 and 2020, costs declined by up to 85 percent, depending on the renewable energy source, making them cheaper than coal and gas in most places.[31] This decline in cost is driven by improved technologies and wider adoption. Since solar capacity gets cheaper as capacity increases, the focus is on increasing energy storage. The price of batteries has fallen by 97 percent in the past 30 years.[32] As I am sure you have noticed, due to government regulation and changes in economics, the demand for electric vehicles (EVs) is also increasing. In fact, in 2021 eV sales doubled from 2020 to 6.6 million, representing the market's sharpest increase.[33]

Even better, companies, countries, and large endowments are shifting away from oil and other fossil fuels. For example, a Dutch court ordered Shell to cut its emissions by 45 percent by the year 2030.[34] Greenland's government announced that it is suspending all new oil and gas exploration activities,[35] and Harvard University, which has the United States' largest academic endowment, announced it will divest from all fossil fuel investments.[36] Further, devastating as Russia's invasion in Ukraine is, it has made more ambitious Europe's energy transition goals. Of course, to make up for the immediate shortfall in energy sources, it has also led to new refining.

Example of progress: Expanding energy access to improve livelihoods in emerging markets

Over the last decade, no greater progress has been made than in the expansion of access to energy in emerging markets, affecting mostly off-grid users in Sub-Saharan Africa. In 2011, roughly 1.2 billion people lacked access to electricity. By 2019, that number had dropped by 900 million people, or from 18 to 9 percent of the world population.[37]

Much of this massive increase in access to energy can be attributed to the important role that catalytic capital and impact investors have played in building the market. Energy-access-related impact investing has seen spectacular growth over the last decade; in fact, according to the Global Impact Investing Network (GIIN), in 2020 it reached approximately $18.6 billion in assets under management (AUM) in emerging markets (EM).[38] That is up nearly eight times from the $2.4 billion AUM reported in 2016.

This is good news, as evidence shows the promising effects of increased access to energy for low-income households. A study[39] executed by Acumen in collaboration with Solar Aid and Google showed that households in Kenya with solar lamps reduce their use of kerosene by about 35 to 40 percent; in addition, because many solar systems include mobile phone charging capacity, these households reduce their need to pay for this service. Furthermore, the household energy ladder points to an even brighter future as many analysts believe that over the next fifteen years, energy consumption will shift from solar lanterns to small home systems and mini-grids.[40] This has already started to happen in East Africa.

Nevertheless, there is opportunity for more progress, and even more investment is needed. Nine hundred million people still lack access to energy, and despite reaching nearly $20 billion AUM, the amount of investment required to close that energy access gap remains far larger than the current amount of capital flowing in to solve the problem.

Impact investing is still far too subscale

To speed progress and achieve the goals the world has collectively set—notably, reaching SDGs by 2030—far more resources will be needed than are currently allocated by governments and other actors. As of 2018, the United Nations estimated that the annual financing gap to achieve the SDGs was $2.5 trillion,[41] yet as a result of the COVID-19 pandemic estimates now suggest that the financing gap to achieve SDGs increased by nearly 50 percent.[42] An additional $3.7 trillion per year must be spent above the estimated $2 trillion that governments currently contribute toward global development challenges (Fig. 1.1).

To put the size of the financing gap in perspective, consider that the annual production of Germany in 2020 was roughly $3.8 trillion.[43] To achieve the SDG goals by 2030, we must add the equivalent of the annual production of Germany to what the world already collectively spends.

On one hand, these numbers seem daunting, especially if you believe (like many do) that governments will be unable to add financing to close this gap and that it will be the responsibility of private actors—corporations, investors, and philanthropists—to help close the annual financing gap. It is hard to believe that this will be possible, given the current rates of spending and charity we see in the world.

There is hope, however. According to the International Finance Corporation (IFC), impact investing in private markets could be as large as $2.1

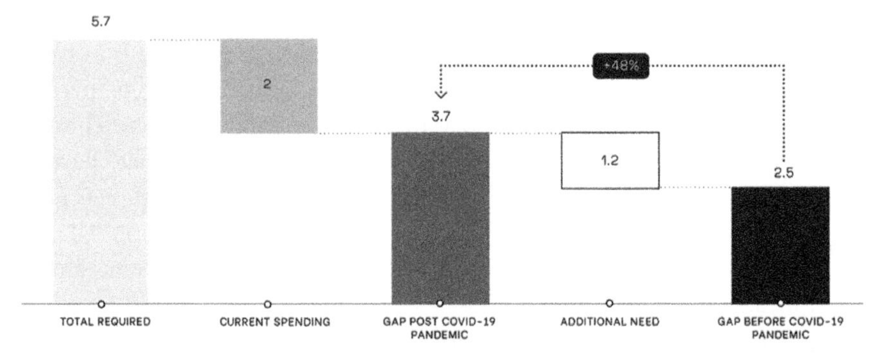

Fig. 1.1 Annual spending required to achieve the SDGs by 2030 ($US trillions)

trillion AUM.[44] This represents significant growth of a field that didn't even exist two decades ago, suggesting progress is possible. Indeed the growth of impact investing in private markets is expected to continue to outpace overall market growth.

Yet when we put a stricter definition of impact, the story becomes more complicated. Data from the Global Impact Investing Network suggest only half of the IFC estimate or $1.1 trillion across 3349 organizations, is clearly measured for impact, including both social/environmental and financial returns.[45] Another and perhaps stricter definition, which looks just at the 152 signatories of the IFC operating principles for impact management[46] that actively manage and seek to "contribute solutions" through their investments, shows the amount of current impact investing drops to $428 billion AUM. Note also, as AUM, these are stock and not flow variables, so it won't be sufficient to simply redirect these assets toward SDG goals. The size of the need is far greater.

Further, when you put impact investing in the context of the larger private wealth and investment management industry, the picture begins to change. The private asset management industry pre-COVID was estimated at $7.3 trillion[47] AUM, nearly 15 times larger than impact investing strategies. The majority was invested in private equity strategies that don't consider their positive contributions to society. The largest strategy within private equity is buyout funds, which actually may cause serious harm to a wide range of stakeholders in the short term. Looking at the broader market and considering both public and private asset management, we see that total AUM jumps to an estimated $103 trillion[48]—more than 200 times larger than current impact investing strategies (see Fig. 1.2).

When we consider that up to one-third of the total asset management industry is, or soon will be, pursuing environmental, social, and governance (ESG) principles, a more promising outlook emerges, with almost $40 trillion in responsibly invested resources. As we will discuss in Chapter 8, however, ESG alone is not enough to make a difference, as many of those investments do not intentionally contribute measurable solutions to SDGs. Significant progress could be made toward achieving SDG goals by diverting more capital away from ESG passive and non-impact intentional approaches and toward impact investing strategies that contribute solutions.

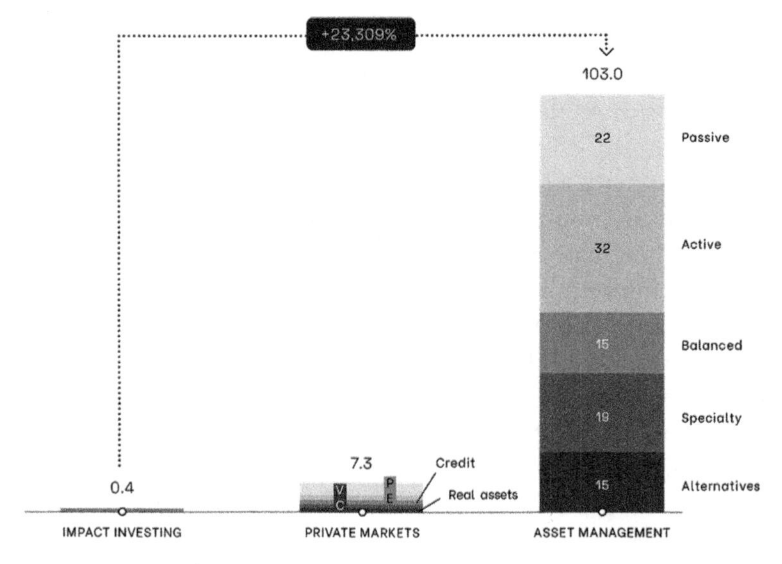

Fig. 1.2 Comparison of impact investing versus other asset management strategies, 2020 (US$ trillions)

Systems can change

The promising news is that the majority of that $100 trillion AUM plus in capital is held by individuals like you and me, individuals with the agency and time to start making changes. The good news is that the next generation of asset owners appears ready to make significant changes in how they invest moving forward. Institutional asset owners such as pension funds, sovereign wealth funds, and insurance companies hold a minority portion of capital, but still represent significant power, as discussed later on.

Returning to individual investors, a growing population of young professionals—both millennials and Gen-Xers—with a net worth < US$1M are increasingly interested in finding ways to align their investments with their values, and to do so sustainably. To gain perspective into their investment and charity donation decisions, as part of writing this book I surveyed nearly a hundred young people under age 45 from across the globe; all were full-time employed, graduates of MBA programs, and affluent professionals with incomes greater than $150,000 per year. What I discovered was comforting. It boiled down to three important insights.

The first insight was that most respondents considered impact investing a viable investment strategy. When respondents were asked about their level of agreement with various statements about impact investing, they produced strong messages. Almost half indicated either agreement or complete agreement that values are an important part of investment decisions and that impact investing is a viable investment strategy. More than 60 percent agreed or strongly agreed that it is possible to intentionally seek and measure social and environmental performance with financial return when making investments, and roughly half disagreed that the only metric useful in determining value creation is financial return.

Second, most respondents have or intend to add impact investments to their portfolio within the next five years. Over 90 percent of the respondents indicated that they either already had or would consider adding impact investments to their personal investment portfolio in the coming five years. From that group, one-third stated they already had impact investments in their portfolio, and another 20 percent said they planned to add an impact investment allocation in the coming year. A whopping 100 percent of the respondents under the age of 35 indicated they had or intended to add impact investments to their personal portfolio over the next five years.

In addition to the individual investors whom I surveyed; this growing interest is also observed among LP respondents to Pitchbook's Sustainable Investment Survey. Of Pitchbook's more than one hundred and fifty LP respondents, 57 percent said they are already making impact investments and another 18 percent said they are developing an impact investing approach.[49]

Finally, in contrast, many potential impact investors lack the knowledge necessary to make impact-related investments. Less than half of the survey respondents indicated they knew how to make an impact-related investment, and 40 percent of those indicated that lack of knowledge was the primary factor limiting them from making such investments. Further, among the more than two dozen impact investment organizations mentioned in the survey, only one (Acumen Fund) was known to more than half of the respondents. The rest of the funds lacked name recognition and offered solutions that varied significantly in terms of type of capital provided, minimum ticket investment requirements, sector of focus, etc. Interestingly, almost none of the organizations offered investment platforms for non-accredited investors. Only three of the more than 25 organizations offered any sort of retail product that would allow individuals with net worth < $US1M to invest in their funds. These insights suggest that many impact investment movements have room to grow by tapping into this expanding base of young, increasingly affluent asset owners.

It clearly will take systems-level change to alter the status quo of how capital is currently allocated and to push individuals and institutions in the direction of positively contributing to solutions that address our planet's most urgent needs.

Capitalism is still the best solution, but we can do better

Some would argue that the system itself—capitalism—is broken, and that to make significant change we need to reimagine or even reduce the role of capitalism itself. This book sets out the reasons why I disagree. Capitalism may not always produce the best outcomes—especially when institutions are too weak to prevent private capture and exploitation of markets—but it is still the best system available among a weak set of alternatives. The challenge that confronts us is to seek ways to make capitalism work better—*to restore integrity to finance and make investing once again about creating value for society as a whole.*

This book is about how we can start making those changes using strategies, structures, and tools that take advantage of capitalism's strengths and how they can work in concert to create systems change. It is about a reimagined financial system that is more inclusive and accountable to all. A shift is needed away from extractive, short-term practices in the name of shareholder primacy and toward a system that values the role of all stakeholders, including workers, communities, and the physical environment.

The overall structure required would adhere to six key strategic paradigm shifts in how we think about traditional finance and investment for development outcomes.

- Seek **financial health,** *not* financial access.
- Provide **patient** capital, *not* venture capital.
- Be an **impact-first investor**, *not* an ESG investor.
- Offer **catalytic finance,** *not* just blended finance.
- Measure success based on **results,** *not* on activities.
- Provide **capacity** building, not just capital.

It is not enough to make only one of these paradigm shifts. These shifts must be made taking a systems approach, recognizing their interdependencies and linkages. In a way which we address changes to the current form of capitalism in concert, as part of a coordinated effort to make it work better for everyone.

Each of these six strategic shifts will be addressed over individual sections, each structured in three chapters: The first of the three will lay out the strategic rationale for why we need to shift the way current practices are implemented; the second will offer my insights into what structures work best in practice, based on nearly two decades of advisory work with leading development finance institutions, family offices, foundations, and private equity funds; and the third will invite a viewpoint from a practitioner who has pioneered a successful financing approach relevant to the paradigm shift. Among the practitioners committed to this project are Sasha Dicther from 60 Decibels, Chris Jurgens from the Omidyar Network, Olivia Prentice from Bridges Ventures and formerly of the Impact Management Project, Joan Larrea from Convergence, Brian Milder from Aceli Africa, Bjoern Struewer from Roots of Impact, and Nicholas Coloff from the Argidius Foundation. By providing these voices, the book will help readers to not only imagine a set of new approaches to redefining the problem they seek to solve but will also redirect their solutions toward using finance and investment as a centerpiece of their efforts to achieve the better world we all desire.

The graphic on the next page provides a general outline of the book overall and is followed by a summary of each of the six major paradigm shift strategies. A conclusion will wrap up by highlighting the key themes and underscoring my thoughts on how to bring real change into practice.

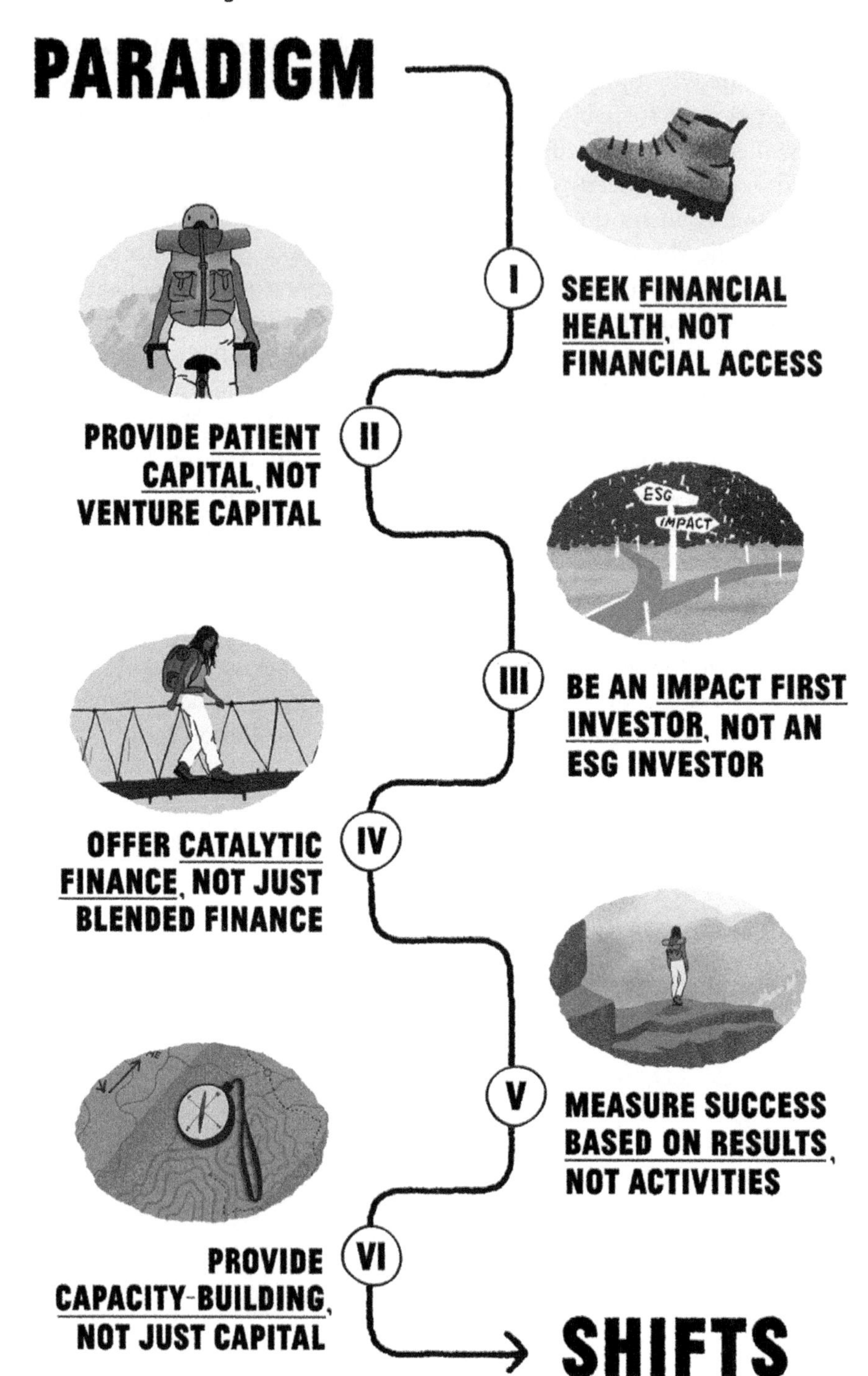
PARADIGM
ESG
IMPACT
I
SEEK FINANCIAL HEALTH, NOT FINANCIAL ACCESS
II
PROVIDE PATIENT CAPITAL, NOT VENTURE CAPITAL
III
BE AN IMPACT FIRST INVESTOR, NOT AN ESG INVESTOR
IV
OFFER CATALYTIC FINANCE, NOT JUST BLENDED FINANCE
V
MEASURE SUCCESS BASED ON RESULTS, NOT ACTIVITIES
VI
PROVIDE CAPACITY-BUILDING, NOT JUST CAPITAL
SHIFTS

Seek **financial health,** *not* financial access (Chapters 2–4).

The first three chapters of the book focus on how commercial banks, fintechs, and other providers of financial services to individual consumers must change their goals. They need to shift from creating access to improve financial outcomes for the hundreds of millions who have been excluded until now to making sure that people can use financial products to better weather economic shocks and invest in health, education, or businesses. It profiles my four years of work with the Bill and Melinda Gates Foundation, the MasterCard Foundation, and CGAP/World Bank Group, which consistently showed that providing access to finance is not enough: To be successful, finance must be considered holistically, offering nonfinancial services bundled with credit or insurance. Also key is listening to consumers, understanding their real needs, and measuring performance from the consumer's perspective. To convey this approach and perspective, the text turns to Sasha Dicther, who shares his insights from building and growing the 60 Decibels/Lean Data approach to impact measurement for financial health.

Provide **patient capital,** *not* venture capital (Chapters 5–7).

This section centers on a fundamental reconceptualization of the venture capital model to align it with the needs of small and growing businesses (SGBs) across the globe. Accessing financing is particularly challenging for certain types of SGBs, such as early-stage ventures and businesses with moderate growth prospects. They are stuck squarely in the "missing middle" of enterprise finance: Too big for microfinance, too small or risky for traditional bank lending, they lack the growth, return, and exit potential sought by venture capitalists. A shift toward a more patient model focused on the true financing needs of these enterprises provides a path forward. The section profiles the work I have done with the Acumen Fund, Omidyar Network, the Dutch Good Growth Fund, and Argidius Foundation, among others, to show how this shift can work in practice. The final chapter brings forward the thoughts and experience of Chris Jurgens from Omidyar Network, who shares how Omidyar's family office puts this strategy into practice.

Be an **impact-first** investor, *not* an ESG investor (Chapters 8–10).

This three-chapter section challenges ESG investing as insufficient to meet the widespread need for change. The proliferation of large private equity firms that raise billion-dollar impact funds and the common practice of greenwashing are problematic. Rather than focus on scale, investing should take an intentional, impact-driven approach, with measurement and accountability, to ensure truly value-oriented impact investing strategies. Drawing on my work with the International Finance Corporation, the United States Development Finance Corporation, BlackRock, and MatterScale Ventures, I lay out the problem and present insights into how some investors overcome this in practice. The section takes a deep dive into the top-of-mind impact investing themes of climate action and racial equity, arguing that funds with these focuses must take an intersectional approach to all of their investing activity. The section closes with insights from Olivia Prentice, formerly of the Impact Management Project, into how that platform has helped bring stakeholder voices front and center into impact investing. In so doing, the Impact Management Project ensured that impact intentionality and measurement do not receive mere lip service but are fully delivered in practice.

Offer **catalytic finance,** *not* just blended finance (Chapters 11–13).

In this part, I turn readers' attention to the role of governments and donors in providing development finance. The chapters argue that the central role of development finance institutions is to push markets to areas experiencing true market failure and lack of adequate finance. More risk-tolerant catalytic capital must be provided and blended with traditional, more return-seeking sources of finance that may help crowd in and scale financing for development goals. Building on my advisory work with Convergence, USAID, and the MacArthur Foundation, among other organizations, I show how this can work in practice with a specific lens toward market failures in healthcare, education, and climate action. Joan Larrea, the CEO of Convergence, continues the discussion, explaining the history and trends in blended finance and presenting the case behind the need for catalytic actors who can truly bring financing for development to scale.

Measure success based on **results,** *not* activities (Chapters 14–16).

This section looks into the practices of innovative, results-based finance—one of the most commonly misunderstood and underutilized tools in all of finance and investment. The approach focuses more on the outcomes of actions and less on whether all activities were completed as stipulated. I start this section by explaining why a more results-centered approach is needed, particularly in government- and donor-funded social programs. I then explore with readers the menu of different approaches to providing results-based finance and why the most popular tool, the impact bond, is not always the best solution, particularly for complex transactions. The section closes with the insights of Bjoern Struewer, from Roots of Impact, who explains the social impact incentives, a simpler tool that can help align incentives and scale up financing for successful programs.

Provide **capacity** building, not just capital (Chapters 17–19).

The final part investigates a unifying theme across the book—the need to combine finance and investment with capacity-building support. While sometimes maligned, offering customized support to entrepreneurs and/or leadership teams plays a critical role in ensuring the impact potential of business activities are realized. I start this section explaining the different types of and delivery models for capacity-building services. I then discuss the evidence on five practices that help to make these types of nonfinancial support activities effective at boosting growth and performance. The section closes with an interview with Nicholas Colloff, the Executive Director of Argidius Foundation, about his learnings of more than two decades funding enterprise support programs in emerging markets.

All of this book's chapters are based on my own advisory work on more than two hundred individual advisory or investment projects with leading development finance institutions, family offices, foundations, and private equity funds. I also lean on insights from a large network of peers and friends who also actively work in the space. I hope these insights resonate, and I look forward to continuing the conversation as markets and uses of capital in society further evolve toward making a better world truly a reality.

Notes

1. Lowder S, Sánchez M, Bertini R (2021) Which Farms Feed the World and Has Farmland Become More Concentrated? *World Development* 142. https://doi.org/10.1016/j.worlddev.2021.105455
2. New York Times (2021) A Vivid View of Extreme Weather: Temperature Records in the U.S. in 2021. https://www.nytimes.com/interactive/2022/01/11/climate/record-temperatures-map-2021.html. Accessed 22 Feb 2022.
3. Centre for Research on the Epidemiology of Disasters, United Nations Office for Disaster Risk Reduction (2020) The Human Cost of Disasters: Overview of the Last 20 years 2000–2019. https://www.undrr.org/publication/human-cost-disasters-overview-last-20-years-2000-2019. Accessed 22 Feb 2022.
4. Pidcock R, McSweeney R (2021) Mapped: How Climate Change Affects Extreme Weather Around the World. Carbon Brief. https://www.carbonbrief.org/mapped-how-climate-change-affects-extreme-weather-around-the-world/. Accessed 22 Feb 2022.
5. UNICEF (2021) The State of the Global Education Crisis: A Path to Recovery. https://www.unicef.org/reports/state-global-education-crisis. Accessed 23 Jun 2022.
6. United States Department of Education (2021) Education in a Pandemic: The Disparate Impacts of COVID-19 on America's Students. https://www2.ed.gov/about/offices/list/ocr/docs/20210608-impacts-of-covid19.pdf. Accessed 23 Jun 2022.
7. United Nations (2020) The Sustainable Development Agenda. https://www.un.org/sustainabledevelopment/development-agenda/. Accessed 23 Jun 2022.

8. Hurtado P (2021) Ex-TPG Star Bill McGlashan Gets Three Months in College Scam. Bloomberg News. https://www.bloomberg.com/news/articles/2021-05-12/ex-tpg-star-bill-mcglashan-gets-three-months-in-college-scandal. Accessed 23 Jun 2022.
9. Clark S, Louch W (2021) *The Key Man: The True Story of How the Global Elite Was Duped by a Capitalist Fairy Tale*. HarperCollins, United States.
10. Nobel Prize (1986) The Sveriges Riksbank Price in Economic Sciences in Memory of Alfred Nobel 1986. https://www.nobelprize.org/prizes/economic-sciences/1986/summary/. Accessed 21 Jun 2022.
11. Kubzansky M (2020) Milton Friedman's Vision Has Failed. Let's Bury It and Move On. Barrons. https://www.barrons.com/articles/reinventing-capitalism-puts-people-over-profits-51600375230. Accessed 23 Jun 2022.
12. Chetty R, Hendren N, Jones M et al (2020) Race and Economic Opportunity in the United States: An Intergenerational Perspective. *Quarterly Journal of Economics* 135(2): 711–783. https://doi.org/10.1093/qje/qjz042
13. Council on Foreign Relations (2022). https://www.cfr.org/backgrounder/us-inequality-debate. Accessed 9 Oct 2022.
14. World Development Indicators (2021) World Bank Group. https://data.worldbank.org/topic/poverty. Accessed 22 Feb 2022.
15. World Bank Group (2022) Extreme Poverty, 2015–2022. https://www.worldbank.org/en/topic/poverty. Accessed 28 Jun 2022.
16. Kim JY (2018) Remarks by World Bank Group President Jim Yong Kim at the 2018 Annual Meetings Plenary. World Bank Group. https://www.worldbank.org/en/news/speech/2018/10/12/remarks-by-world-bank-group-president-jim-yong-kim-at-the-2018-annual-meetings-plenary. Accessed 22 Feb 2022.
17. Chancel L, Piketty T, Saez E et al (2022) World Inequality Report 2022. World Inequality Lab.
18. World Economic Forum (2021) Global Gender Gap Report 2021.
19. Gender Wage Gap Data (2020) OECD. https://www.oecd.org/gender/data/employment/. Accessed 22 Feb 2022.
20. World Economic Forum (2021) Global Gender Gap Report 2021.
21. Collins C, Asante-Muhammed D, Hoxie J et al (2019) *Dreams Deferred: How Enriching the 1% Widens the Racial Wealth Divide*. Institute for Policy Studies.
22. Collins C, Asante-Muhammed D, Hoxie J et al (2019) *Dreams Deferred: How Enriching the 1% Widens the Racial Wealth Divide*. Institute for Policy Studies.
23. World Population Prospects (2019) UN Population Division. https://population.un.org/wpp/. Accessed 22 Feb 2022.
24. WHO, UNICEF, UNFPA, World Bank Group, UN Population Division (2015) Trends in Maternal Mortality: 1990 to 2015.

25. Many basic health and nutrition indicators have deteriorated globally since the COVID-19 pandemic but precise levels have not yet been estimated at the time of the writing of this book.
26. FAO, IFAD, UNICEF, WFP, WHO (2021) The State of Food Security and Nutrition in the World 2021. https://doi.org/10.4060/cb4474en
27. UNAIDS (2021) Global HIV & AIDS Statistics – Fact Sheet. https://www.unaids.org/en/resources/fact-sheet. Accessed 22 Feb 2022.
28. WHO (2021) World Malaria Report 2021.
29. Climate Action Tracker (2021) Glasgow's 2030 Credibility Gap: Net Zero's Lip Service to Climate Action. https://climateactiontracker.org/publications/glasgows-2030-credibility-gap-net-zeros-lip-service-to-climate-action/. Accessed 28 Jun 2022.
30. World Meteorological Organization (2021) Atlas of Mortality and Economic Losses from Weather, Climate and Water Extremes (1970–2019).
31. International Renewable Energy Agency (2021) Renewable Power Generation Costs in 2020.
32. Ziegler M, Trancik J (2021) Re-examining Rates of Lithium-Ion Battery Technology Improvement and Cost Decline. *Energy & Environmental Science* 14: 1635–1651. https://doi.org/10.1039/D0EE02681F
33. International Energy Agency (2022) Global Electric Vehicle Outlook 2022.
34. BBC News (2021) Shell: Netherlands Court Orders Oil Giant to Cut Emissions. https://www.bbc.com/news/world-europe-57257982. Accessed 22 Feb 2022.
35. Buttler M (2021) Greenland Scraps All Future Oil Exploration on Climate Concerns. Bloomberg News. https://www.bloomberg.com/news/articles/2021-07-16/greenland-scraps-all-future-oil-exploration-on-climate-concerns#xj4y7vzkg. Accessed 22 Feb 2022.
36. Goodman J, Griffin K (2021) Harvard Will Move to Divest Its Endowment from Fossil Fuels. The Harvard Crimson. https://www.thecrimson.com/article/2021/9/10/divest-declares-victory/. Accessed 22 Feb 2022.
37. World Development Indicators (2020) World Bank Group. https://data.worldbank.org/indicator/EG.ELC.ACCS.ZS. Accessed 22 Feb 2022.
38. Hand D, Dithrich H, Sunderji S et al (2020) *2020 Annual Impact Investor Survey*. Global Impact Investing Network.
39. Rom A, Günther I, Harrison K (2017) *Economic Impact of Solar Lighting: Results from a Randomised Field Experience in RURAL KEnya.* Acumen. https://acumen.org/wp-content/uploads/2015/10/Report-The-Economic-Impact-of-Solar-Lighting.pdf. Accessed 24 Jun 2022.
40. Craine S, Mills E, Guay J (2014) *Clean Energy Services for All: Financing Universal Electrification.* Sierra Club.
41. UNSDG (2018) Unlocking SDG Financing: Findings from Early Adopters.
42. OECD (2022) Closing the SDG Financing Gap in COVID-19 Era. https://www.oecd.org/dev/OECD-UNDP-Scoping-Note-Closing-SDG-Financing-Gap-COVID-19-era.pdf. Accessed 09 Sept 2022.

43. World Development Indicators (2020) World Bank Group. https://data.worldbank.org/indicator/NY.GDP.MKTP.CD?locations=DE. Accessed 28 Jun 2022.
44. International Finance Corporation (2020) Growing Impact: New Insights into the Practice of Impact Investing.
45. Hand D, Ringel B, Danel A (2022) *Sizing the Impact Investing Market: 2022*. The Global Impact Investing Network (GIIN), New York.
46. Operating Principles for Impact Management (2022) Signatories and Reporting. https://www.impactprinciples.org/signatories-reporting. Accessed 24 Jun 2022.
47. McKinsey & Company (2021) McKinsey Global Private Markets Review 2021: A Year of Disruption in the Private Markets.
48. Heredia L, Bartletta S, Carrubba J et al (2021) *The $100 Trillion Machine, Global Asset Management 2021*. Boston Consulting Group.
49. Pitchbook (2021) 2021 Sustainable Investment Survey.

Part I

Seek Financial Health, Not Financial Access

2

Reassess the Financial Inclusion Revolution

Financial intermediaries—whether banks, microfinance institutions, nonbank financial intermediaries (NBFIs), insurance providers, or fintechs—are deeply enmeshed in our lives. They provide credit, savings, insurance, and payment services to the growing population of consumers and businesses across the globe. In fact, the global financial services market size reached $23.3 trillion in 2021 and is expected to continue to grow rapidly.[2] There are more than 10,000 regulated commercial banks in the United States and Europe Union and more than 10,000 microfinance institutions in emerging markets, according to the ECB, the FDIC, and MIX Market data. A recent study shows more than 25,000 fintech start-ups across the globe, most heavily concentrated in emerging markets.[3] With this growing ubiquity of financial services, it seems as though significant progress has been made toward broad access. Yet access to financial services isn't all that matters.

I can remember the first time I applied to and received a credit card as a college freshman. In high school my father had forbidden me from having access to any sort of credit card, and only recently had he allowed me to have a debit card linked to a checking account associated with my meager savings. Suddenly as an adult and a college freshman I discovered I didn't need my dad to have access to credit and decided upon a tip from a friend to build my credit rating by taking out as many credit cards under my name as I could find. I didn't plan to use any of them but I thought this was a good idea, as a strong credit rating seemed like something I might want to have in the future. As a consequence, I improved my financial access to new products—in this

K. Hornberger, *Scaling Impact*,
https://doi.org/10.1007/978-3-031-22614-4_2

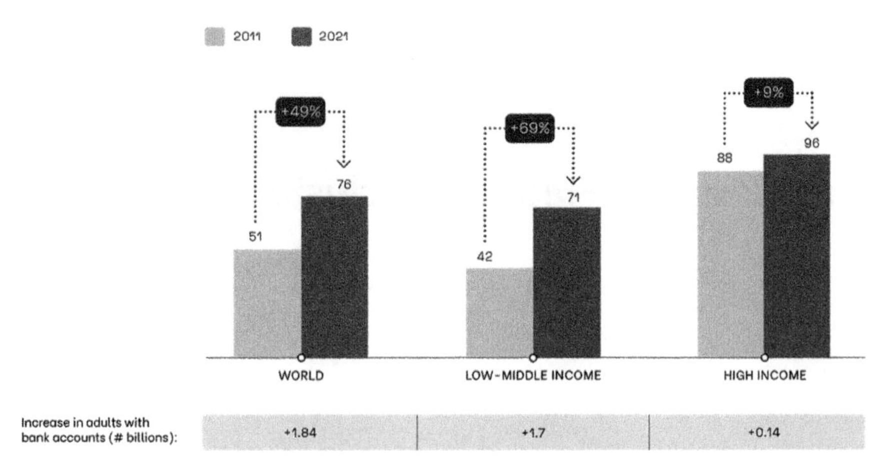

Fig. 2.1 Change in share of adult population with bank accounts (% ages 15+), 2011–2021 (*Source* World Bank Group, Global Findex Database, Author analysis)

case a credit card—but I didn't improve my livelihood, since I didn't use the credit cards and didn't, as I should have, even pay attention to whether there were any risks involved. Instead, I continued to rely on a small income from a college campus job and my debit card to pay for my out-of-pocket expenses.

My story is not that dissimilar to many across the world. Many people gain access to financial services, but, like me, they may not always understand how to use them, how they fit into our lives, what benefits they offer, or if they are designed for and suitable to our needs. Sadly, in many cases, people without a financial backstop, such as my parents provided me (I had at least a half dozen overdraft warnings on my checking account in college), wind up getting into debt that can have material financial consequences that negatively impact their livelihoods and financial well-being.

Over the last decade the push for financial inclusion (that is, access to financial services for people who were previously "unbanked") has united corporates, entrepreneurs, governments, and social sector leaders—and with remarkable results. In 2011, only 51 percent of the world's adults had a formal bank account. By 2021, as the World Bank recently reported in its new Global Findex data, we've reached 76 percent: 1.8 billion more people are now connected to modern financial services, of which 1.7 billion are coming from low- and middle-income economies (see Fig. 2.1).[4]

The COVID-19 pandemic has accelerated this trend, particularly via a shift toward digital financial services. Consequently, more people in emerging markets and low-income populations in developed markets have gained access to the financial system than ever before. According to GSMA, in 2021, the value of mobile money transactions in low- and middle-income countries (LMICs) reached the trillion-dollar mark.[5]

Yet much like my own experience as a college student, a large cohort of people across the world are not well served by the existing financial system. Many, like I once did, have open credit accounts but do not use them. Others have incomes that are low or irregular, and thus they rely on expensive solutions such as informal money lenders, payday loans or check-cashing services to lead their financial lives. The reality is that access alone does not truly solve people's financial struggles or set them up for long-term success.

Financial access is not the same as financial health.[6] Greater access to financial services matters less than whether your financial services contribute to improving your life. Financial health is about whether people are spending, saving, borrowing, and planning in a way that will enable them to be resilient and pursue opportunities that improve their lives over time.[7] On this point, the evidence remains mixed and nuanced. For example, the latest comprehensive literature review from more than twenty years of randomized control trials looking at the impact of microcredit on the livelihoods of the poor demonstrated that microcredit was neither a panacea against poverty nor a harmful intervention: more generally, it was found to create modest but neither revolutionary nor deleterious impacts.[8] Even where evidence is relatively strong (as it is, for example, on the impact of savings on investment), it is important to ask both what outcomes are attributable to financial services and what is the context of the accumulated evidence. Can we extrapolate conclusions based on one place to another?[9]

Improving financial health is also about ensuring that financial services are affordable and fit with people's financial needs. Individuals clearly differ in age, gender, income levels, and attitudes toward risk. Small business owners differ in their ambitions, growth trajectories, and financing needs. What works for a small-plot farmer in Guatemala is not what works for an auto-repair shop owner in Rwanda.

Finally, we need to remember that financial inclusion is merely a means to an end. Access to traditional financial products does not automatically result in improved financial stability, resilience, or access to opportunity. Many financial products were not developed with these pragmatic goals in mind.[10] Rather than financial inclusion, the goal should be financial health that allows everyone to manage day-to-day expenses, have resilience against unexpected economic shocks, and plan effectively for the future to achieve progress in life.[11] By providing access to financial services we are not actually creating financial health in people's lives unless those services are used in the right way, toward the right outcomes, to meet individuals' needs.

Nowhere is this truer than in the growing fintech ecosystem, a space the *Economist* magazine predicted would reshape finance.[12] No question fintech is growing and maturing in the world's largest economies. In the United States, the share of unsecured personal loans obtained via fintech increased from 5 percent in 2013 to 39 percent in 2018.[13] In China in 2019, customer adoption rates for fintech payments and banking services were estimated at greater than 90 percent.[14] And according to CB Insights, in June 2022, 243 of the world's 1170 unicorns (private companies valued over US$1 billion) were fintechs.[15]

This dynamic growth has generated the belief that the fintech revolution is remaking emerging and frontier markets. Many observers believe that fintech models will succeed to scale and serve the base of the pyramid in ways that microfinance never could. Despite reasonable hypotheses and some evidence about its shape, the nature of the market, and its customers and impact, fintech requires a closer look before conclusions are drawn about its future.

Hypothesis 1: Fintech serves a large total addressable market (TAM) in emerging markets

Validated. My back-of-the-envelope calculations, made using publicly available World Bank and IMF data for six large emerging markets in Africa, Asia, and Latin America, indicate that those six markets alone represent at least a $741 billion market (see Fig. 2.2)—without even considering unmet financing needs estimated in the trillions of dollars in previous IFC studies.[16]

Hypothesis 2: Fintechs are growing faster than traditional financial services segments

Validated. The mobile money industry (as a proxy for fintech overall) has more than doubled in value since 2017, a pace of growth much faster than that of traditional banking services, and now processes more than $2 billion a day.[17] For example, comparing the growth of borrowers and depositors at commercial banks with the growth of registered mobile money accounts in Latin America and the Caribbean, we can see that mobile money has been growing by 18 percent a year, outpacing growth of traditional banking services (see Fig. 2.3). The same pattern appears in other emerging market regions.

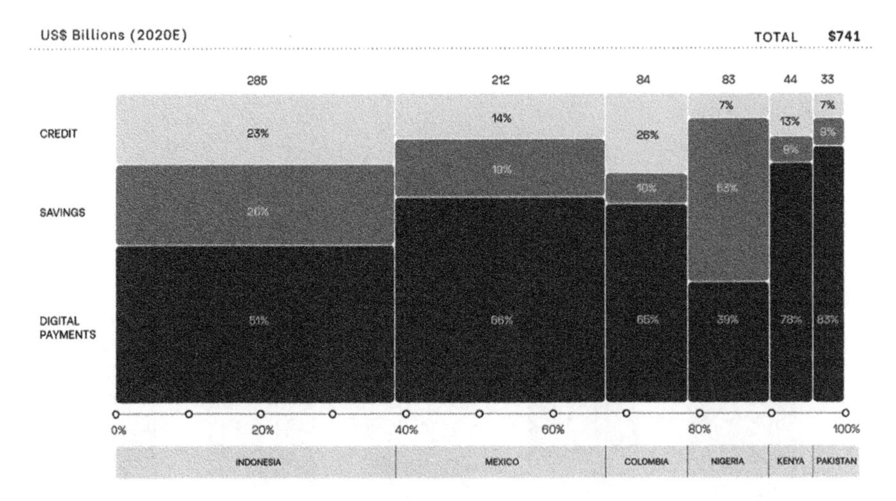

Fig. 2.2 Estimate of micro and small enterprise (MSE) financial services market size in six emerging markets (*Source* IMF, World Bank Group, Global Findex Database, Author analysis)

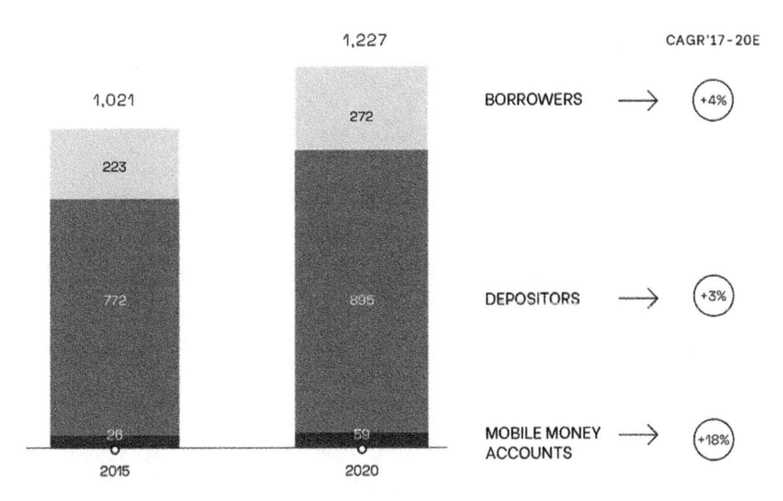

Fig. 2.3 Growth of borrowers, depositors, and mobile money accounts in Latin America and Caribbean, number per 1000 adults, 2015–2020 (*Source* World Bank Global Development Indicators, GSMA State of the Sector Report)

Hypothesis 3: Fintechs attract even greater investor interest due to their dynamic growth

Validated. Fintech venture capital and private equity have grown explosively. According to Crunchbase data, over the last five years fintech equity investments (excluding follow-on rounds) totaled approximately $176 billion, with

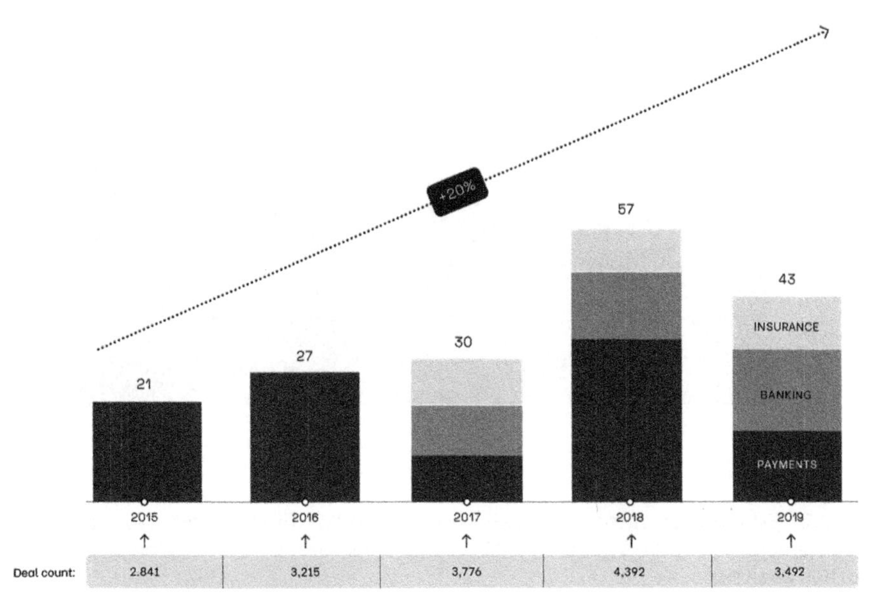

Fig. 2.4 Total invested in fintech companies by value and deal count, US$ billions, 2015–2019 (*Source* Crunchbase, Author analysis)

20 percent year-on-year growth (see Fig. 2.4).[18] As of June 2022, a whopping 21 percent of the current global unicorns are fintech companies, including some in emerging markets, such as Nubank, Ovo, and Flutterwave.[19]

As fintech investments have been growing, microfinance investing has slowed, despite having been touted as a "silver bullet" for financial inclusion. Data from Symbiotics' surveys show that from 2015 to 2019 the number of microfinance investment vehicles has remained flat and total AUM managed by these funds has increased by only 10 percent year on year.[20] Further, many traditional impact-oriented microfinance investors, like Bamboo Capital, Accion, and even the IFC, have shifted toward building specialized fintech investment portfolios or funds, such as Quona and Flourish.

Hypothesis 4: Fintechs serve base-of-pyramid populations and small businesses

Not enough evidence. While fintechs hold promise for meeting the needs of the base-of-the-pyramid population in the bottom one-third of income, they are not yet doing so at scale. In 2021, World Bank data showed the poorest 40 percent of residents in low- and middle-income countries were more than a third less likely than others to use digital payments (Fig. 2.5).[21]

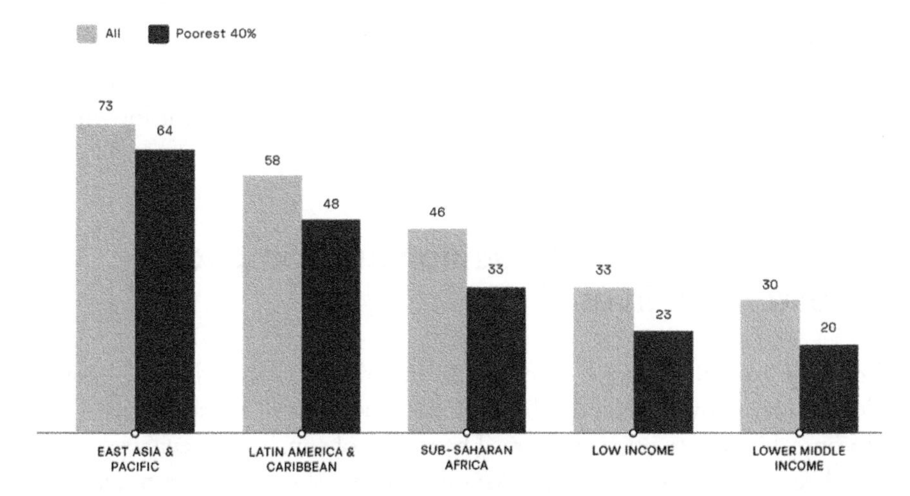

Fig. 2.5 Comparison of usage rates of digital payments by region and income status, 2021 (*Source* World Bank, Global Findex Database, Author analysis)

Further, to please the venture and growth investors pumping money into them, many fintech companies attempt to prove their business models work by serving middle- and high-income customers in urban settings and do not have specific financial inclusion mandates as part of their business model. As such, many fintech companies do not design products to meet low-income populations' needs, such as affordability.

Hypothesis 5: Fintech products are superior to those of traditional finance providers

Unclear and potentially harmful. Without doubt, fintechs hold promise for closing financing gaps by digitizing or systematizing services to lower per-transaction costs to serve. However, digital finance without customization or informed service representatives to explain its uses and benefits may lead to suboptimal outcomes at best and to default at worst. In addition, financial services work best when accompanied by nonfinancial services, such as agronomic technical assistance or business coaching. To date, digital models that provide coaching and advisory support have not yet succeeded in completely replacing higher-cost, in-person services or in proving they can provide higher-quality delivery and results.

Further, fintech business models also introduce new customer risks and questions that to date have not been fully addressed. For example, who owns the smartphone, phone numbers, or data associated with the account, and what happens if or when they are stolen? And is fintech's rise just a shift from financial services provided by large commercial banks to services provided by large technology platforms, neither of which are designed for low-income populations or small businesses?[22]

Hypothesis 6: Fintechs demonstrate impact on customers' lives

Too early to tell. Relative to other means of financial inclusion—notably microfinance's impact on the lives of the poor—little research has looked into fintech's impact on customers. The results of some peer-reviewed studies have been promising, however. Mobile money shows strong potential to improve customers' savings behavior and welfare (see Fig. 2.6).[23] One study shows digital credit users display healthy borrowing behavior; improved physical, social, and emotional welfare; and investments in income-generating activities. Other evidence, however, shows counterproductive borrowing behavior and deteriorating incomes and welfare. More recently, an NBER working paper on China showed fintech lenders can be predatory, leading to higher levels of customer default and over-indebtedness relative to traditional lending programs.[24]

To synthesize, fintechs show unparalleled potential to help solve long-standing financial inclusion challenges in emerging markets. They are, without question, the future. Much remains to be proven concerning fintech's ability to serve previously unserved or underserved populations, however.

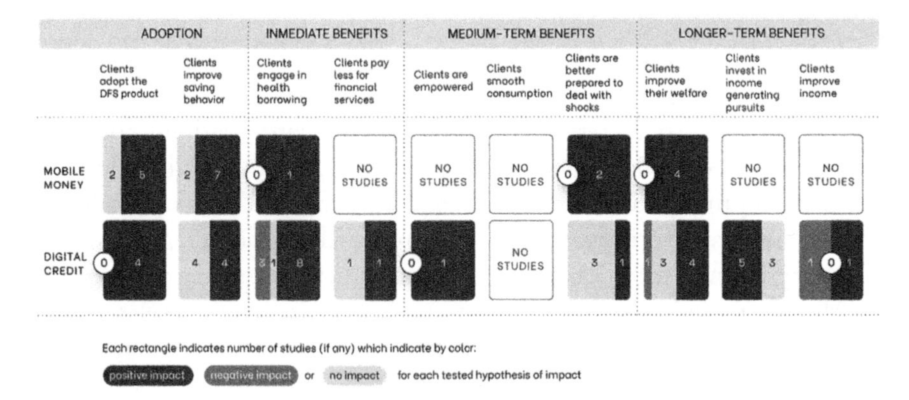

Fig. 2.6 Literature review of peer-reviewed studies testing various benefits of mobile money and digital credit (*Source* MasterCard Foundation, Evidence Gap Map [2019])

Providing superior products and converting growth into impact will be key. From where I sit right now, I draw conclusions similar to those of CGAP.[25] Fintech looks a lot like the microfinance investment space did two decades ago, and as we now know, the promise that microfinance would achieve financial inclusion for emerging markets was more hype than reality.[26] Funders and investors would be wise to stop assuming that fintech is supporting low-income microentrepreneurs and small businesses in emerging markets and should instead seek ways to push these players toward greater inclusivity by valuing quality of services and development outcomes as much as they value growth. Funders and investors would also be wise to invest in more women founders as the average share of fintechs with women founders has hovered around 10 to 15 percent over the last 20 years.[27]

The next chapter explores several ways that financial intermediaries are making the shift from promoting financial access to improving the financial health of their customers.

Summary of key messages from this chapter

- Financial access does not mean financial health.
- The fintech revolution, while a rapidly growing market, has much to prove before we can really rely on it to revolutionize financial services.
- We need to care as much about finding evidence that financial services are materially improving people's lives as we do about whether they merely have access to them.
- We need to design financial products to be affordable and address consumers' needs.
- We should consider bundling financial services with nonfinancial services to maximize the potential that consumers will achieve the outcomes they desire.

Notes

1. The Business Research Company (2022) Financial Services Global Market Report 2022. https://www.globenewswire.com/news-release/2022/05/18/2445691/0/en/Financial-Services-Global-Market-Report-2022.html. Accessed 30 Jun 2022.
2. Ibid.
3. BCG (2022) Fintech Control Tower. https://www.fct.bcg.com/#/home. Accessed 30 Jun 2022.

4. Demirgüç-Kunt A, Klapper L, Singer D et al (2022) The Global Findex Database 2021: Financial Inclusion, Digital Payments, and Resilience in the Age of COVID-19. World Bank Group. https://www.worldbank.org/en/publication/globalfindex. Accessed 30 Jun 2022.
5. Awanis A, Lowe C, Andersson-Manjang S et al (2022) State of the Industry Report on Mobile Money 2022. GSMA. https://www.gsma.com/sotir/. Accessed 30 Jun 2022.
6. Ehrbeck T (2018) Moving from Financial Access to Health. World Bank Blogs. https://blogs.worldbank.org/allaboutfinance/moving-financial-access-health. Accessed 30 Jun 2022.
7. Punatar P, Seltzer Y (2020) Place Financial Health at the Heart of Financial Institutions to Create Business and Customer Value. Accion. https://www.accion.org/placing-financial-health-at-the-heart-of-financial-institutions-to-create-business-and-customer-value. Accessed 30 Jun 2022.
8. Bauchet J, Marshall C, Starita L et al (2011) Latest Findings from Randomized Evaluations of Microfinance. CGAP. https://www.povertyactionlab.org/sites/default/files/publication/FORUM2.pdf. Accessed 30 Jun 2022.
9. World Economic Forum (2018) Advancing Financial Inclusion Metrics: Shifting from Access to Economic Empowerment. https://www3.weforum.org/docs/WEF_White_Paper_Advancing_Financial_Inclusion_Metrics.pdf. Accessed 30 Jun 2022.
10. Punatar P, Seltzer Y (2020) Place Financial Health at the Heart of Financial Institutions to Create Business and Customer Value. Accion. https://www.accion.org/placing-financial-health-at-the-heart-of-financial-institutions-to-create-business-and-customer-value. Accessed 30 Jun 2022.
11. World Economic Forum (2018) Advancing Financial Inclusion Metrics: Shifting from Access to Economic Empowerment. https://www3.weforum.org/docs/WEF_White_Paper_Advancing_Financial_Inclusion_Metrics.pdf
12. The Economist (2020) How the Digital Surge Will Reshape Finance. https://www.economist.com/finance-and-economics/2020/10/08/how-the-digital-surge-will-reshape-finance. Accessed 30 Jun 2022.
13. Transunion (2019) FinTechs Continue to Drive Personal Loan Growth. https://newsroom.transunion.com/fintechs-continue-to-drive-personal-loans-to-record-levels/. Accessed 30 Jun 2022.
14. EY (2019) Global FinTech Adoption Index 2019. https://www.ey.com/en_us/ey-global-fintech-adoption-index. Accessed 30 Jun 2022.
15. CB Insights (2022) The Complete List of Unicorn Companies. https://www.cbinsights.com/research-unicorn-companies. Accessed 1 Jul 2022.
16. IFC (2017) MSME Finance Gap. https://www.smefinanceforum.org/data-sites/msme-finance-gap. Accessed 1 Jul 2022.
17. Awanis A, Lowe C, Andersson-Manjang S et al (2022) State of the Industry Report on Mobile Money 2022. GSMA. https://www.gsma.com/sotir/. Accessed 30 Jun 2022.

18. Crunchbase (2020) Industry Spotlight: Fintech. https://about.crunchbase.com/fintech-industry-report-2020/. Accessed 1 Jul 2022.
19. CB Insights (2022) The Complete List of Unicorn Companies. https://www.cbinsights.com/research-unicorn-companies. Accessed 1 Jul 2022.
20. Symbiotics (2019) 2019 Symbiotics MIV Survey. https://symbioticsgroup.com/publications/2019-symbiotics-microfinance-investment-vehicles-miv-survey/. Accessed 1 Jul 2022.
21. Demirgüç-Kunt A, Klapper L, Singer D et al (2022) The Global Findex Database 2021: Financial Inclusion, Digital Payments, and Resilience in the Age of COVID-19. World Bank Group. https://www.worldbank.org/en/publication/globalfindex. Accessed 30 Jun 2022.
22. Dalberg (2019) Bridging the Credit Gap for Micro and Small Enterprises through Digitally Enabled Financing Models. https://www.findevgateway.org/sites/default/files/publications/files/external_190131_final_report_mses_cgap_external_final_updated-bisvb.pdf. Accessed 11 Jul 2022.
23. MasterCard Foundation Partnership for Finance in a Digital Africa (2019). Evidence Gap Map. https://egm.financedigitalafrica.org/. Accessed 11 Jul 2022.
24. Di Maggio M, Yao V (2020) Fintech Borrowers: Lax Screening or Cream-Skimming? (NBER Working Paper Series, Working Paper 28021). https://www.nber.org/system/files/working_papers/w28021/w28021.pdf. Accessed 11 Jul 2022.
25. Bull G (2019) Great Expectations: Fintech and the Poor. CGAP Leadership Essay Series. https://www.cgap.org/blog/great-expectations-fintech-and-poor. Accessed 11 Jul 2022.
26. Murthy G, Fernandez-Vidal M, Faz X et al (2019) Fintechs and Financial Inclusion: Looking Past the Hype and Exploring Their Potential. CGAP. https://www.cgap.org/sites/default/files/publications/2019_05_Focus_Note_Fintech_and_Financial_Inclusion_1_0.pdf. Accessed 11 Jul 2022.
27. Khera P, Ogawa S, Sahay R and Vasishth M (2022) Women in Fintech: As Leaders and Users. IMF. https://www.imf.org/-/media/Files/Publications/WP/2022/English/wpiea2022140-print-pdf.ashx. Accessed 16 Oct 2022.

3

Design Services That Build Financial Health

Several years ago, when I had the opportunity to speak to entrepreneurs around the globe for a research project I was working on with a large foundation, I met Ana Penilla, a microentrepreneur based in Lima, Peru. The focus of my research was to learn about the financing journey she had undertaken for her small business and about the challenges she had faced in accessing the right kind of financing to help her grow the business. We also discussed the impact that financing had on her livelihood. What I learned was revealing but also consistent with stories I had heard from microentrepreneurs elsewhere through my travels and work.

Ana had started a barber shop in the Miraflores neighborhood that provided traditional and creative men's haircuts, beard trims, and shaves. Initially unable to get a bank loan, she was forced to self-finance to start the business. Fortunately for Ana she had built up enough savings from the wages she had earned at the full-time job she had decided to leave. After demonstrating revenue, she was able to secure a short-term $20,000 bank loan at a 12 percent per month interest rate, using her home as collateral, to purchase assets such as chairs, tables, mirrors, and televisions for the shop. This loan fueled her business's growth, and she nearly doubled her customer base and revenues over the next two years.

A couple of years later, when Ana went to look for more financing to fuel additional growth, the economy had soured and she couldn't get the banks to provide her more working capital or expansion finance: they were looking for "less risky" lending opportunities. Seeing no way out and facing rising

K. Hornberger, *Scaling Impact*,
https://doi.org/10.1007/978-3-031-22614-4_3

costs at home and in the business, she decided her best option was to sell the business and move to another neighborhood. Luckily she was able to find a buyer, but she sold the business below the value of the assets she had bought for $12,000. Ana has now moved to a new neighborhood and is looking to restart her life with a new business.

Ana, like many others, had access to finance, but she found it difficult to access and expensive, it wasn't tailored to her needs, and it wasn't available when she needed it most. Ultimately, instead of thriving and seeing her business grow, Ana failed, and her livelihood and social and psychological welfare deteriorated. But from the lender's perspective, her story was seen as a success, as the revenue of her business improved during the life of the loan. The bank did not care about the affordability and outcomes of the financing it provided. Rather than thinking about Ana's financial health as the goal, the bank that served her, like many around the globe, thought only about her access to finance.

Financial intermediaries of all types, not just banks, need to do better. They should not be in the business of keeping people sustainably poor. To achieve financial health—not just financial access—we need to think about both affordability and design of financial products and services as well as how using financial services leads to improved outcomes and greater economic resilience for customers.

From practical experience advising dozens of clients on themes of financial inclusion across the globe and from having studied and read much of the academic and development research on what works, I have concluded that while there are no silver bullets for financial inclusion that leads to consistently better outcomes of improved livelihoods, resiliency, and dignity for customers, some good practices have proven to work better than others. Specifically, the five practices detailed below are effective and repeatable ways that financial intermediaries can boost the financial health of borrowers.

Segment to understand customer value

Financial intermediaries need to grow to survive, thus customer acquisition is paramount. To fulfill their growth aspirations and gain traction quickly, financial intermediaries have heavily invested in consumer research and human-centered design to understand their customers' needs. However, deep quantitative knowledge on customer segmentation—and its links to customer profitability—continues to be a major challenge for many financial intermediaries.[1] Most financial intermediaries have limited understanding of the economic value of different customer segments and what customer cash flows look like. Accordingly, they generally do not have systems in place to distinguish between different value-segments, the revenue each brings in, and whether the institution can effectively invest in acquiring them.

Financial intermediaries should seek to grow with the right customers. Understanding differences in customer value by segment is fundamental for sustainable growth. Customer lifetime value—defined as the net present profit a financial intermediary will get from serving a customer over its lifetime—can vary widely across segments. By understanding those differences in lifetime revenues and servicing costs between different customer segments, financial service providers can identify specific strategies that support overall profitability and determine how much they should spend on customer acquisition. This knowledge can also be used to target high-potential customers intentionally and successfully.

Most importantly, understanding customer lifetime value can help financial intermediaries put into perspective how much they should spend on customer acquisition.[2] If the lifetime value for a specific customer segment is below its customer acquisition cost, the provider should reflect on whether long-term subsidy is needed or whether a short-term subsidy can be accompanied by other support to help customers graduate to a more profitable segment. (Subsidies should last at least until customer acquisition costs can be lowered.) With a strong pressure for growth, many financial intermediaries may be tempted to burn large amounts of cash on customer acquisition. While spending too little is not advisable, spending too much just to grow at any cost is also dangerous. This delicate balance requires continual monitoring of customer economics and the costs of growth.

One financial intermediary that does this well is VisionFund Myanmar (VFM). VFM is one of the 28 microfinance institutions that form VisionFund. VisionFund is the financial inclusion services provider associated with World Vision, the global Christian relief, development, and advocacy organization. I learned about their work in Myanmar through consultations with VisionFund's Most Missing Middles[3] program manager. The Most Missing Middles program was designed based on segmentation work that VisionFund was doing at the global level.

Through portfolio unit economic analysis, VisionFund staff identified that a large and growing cohort of their customers were seeking larger loans with differentiated services. As such, VisionFund decided to undergo a pilot with VFM to understand the economic value and opportunity of serving a new customer segment. The pilot was successful. A program to provide small and growing businesses with working capital loans between $3000 and $10,000 has now been rolled out and added to VisionFund's traditional group lending offerings.

Prioritize underserved segments, such as female clients

Historically, female clients have been disproportionately underserved by financial intermediaries for a wide range of reasons. Female customers tend to have fewer and/or less access to productive assets.[4] Females also tend to be less mobile and more time-poor than men, to have lower digital literacy, and to have less agency in decision-making. These trends have been exacerbated by the COVID-19 pandemic.

Perhaps unsurprisingly, these factors have historically made financial intermediaries consider females to be less attractive as customers. Females' lower incomes and less access to collateral limit loan sizes available to them. Additionally, providing tailored services and marketing to women has tended to increase the cost to serve. Combining smaller loan sizes and a higher cost to serve often translates into lower customer lifetime value relative to male counterparts.

Serving female clients may be less profitable for some financial intermediaries today, but that could change into the future if structural challenges are addressed. Importantly, women appear to exhibit lower rates of churn and lower default rates than men, presenting opportunities for bigger loan sizes. Assuming that targeted initiatives could enhance female borrowers' incomes through tailored service and financing, large loan sizes could increase the customer lifetime value of their female customers significantly.

Females represent an opportunity to grow the customer base while improving financial sustainability over time. Capturing this opportunity takes time, focus, and deliberate investment in both understanding unique customer segments and designing products and processes that target these segments. In-depth knowledge on digital use patterns, demographics, behavior, and psychology can also help financial intermediaries design appropriate delivery channels, lending processes, and digital interfaces. These investments will, of course, need to be accompanied by better gender representation in management and governance. Finally, even if the market opportunity exists, expanding reach to these customer segments will likely require donor support in the short- and medium-term to offset the initial cost and risk for providers.

A great example of an organization that successfully targets underserved female clients is Friendship Bridge. Friendship Bridge is a Guatemalan microfinance institution (MFI) that offers credit and additional services solely to women borrowers, primarily in indigenous rural areas of the country. From 1990 to date, it has provided nearly 30 thousand female clients per year with loans and provided access to health services for another 13 thousand female clients, the majority of whom previously didn't have access to finance or health services. And Friendship Bridge accomplished all this while delivering superior financial results.[5]

Be holistic

Financial intermediaries can intentionally design holistic value propositions and partnerships to drive product traction and share cost and risk. Holistic value propositions provide comprehensive access to a range of services that help customers address multiple challenges. A complete package of services may be the only way to overcome the vicious cycle whereby credit is required to invest in productivity, but increased productivity and income are needed to access credit. From a business model perspective, my work with the Rural Agricultural Finance Learning Lab and the Sustainable Trade Initiative (IDH) suggests that, in some circumstances, overall financial intermediary profitability could markedly increase if providers offer holistic value propositions.[6]

Over the last decade, holistic value propositions have surged in popularity. In contrast to more traditional credit-only service propositions, holistic value propositions provide or facilitate access to a variety of financial and nonfinancial services. The latter may include agronomic advisory and technology services, access to inputs, or market linkages. It may also include things like access to health insurance or direct health services. These financial and nonfinancial services can be offered as bundles or from a menu of services from which the customer can choose. Furthermore, these services can be offered by a single actor or a coalition of actors. Figure 3.1 shows an example of a holistic value proposition carried out by a coalition of actors serving smallholder farmers.

From the customers' perspective, such holistic value propositions are often more attractive than finance-only products. Most customers face significant challenges beyond access to finance. Holistic value propositions provide comprehensive access to a range of services that help customers address multiple challenges. For many, a complete package of services may be the

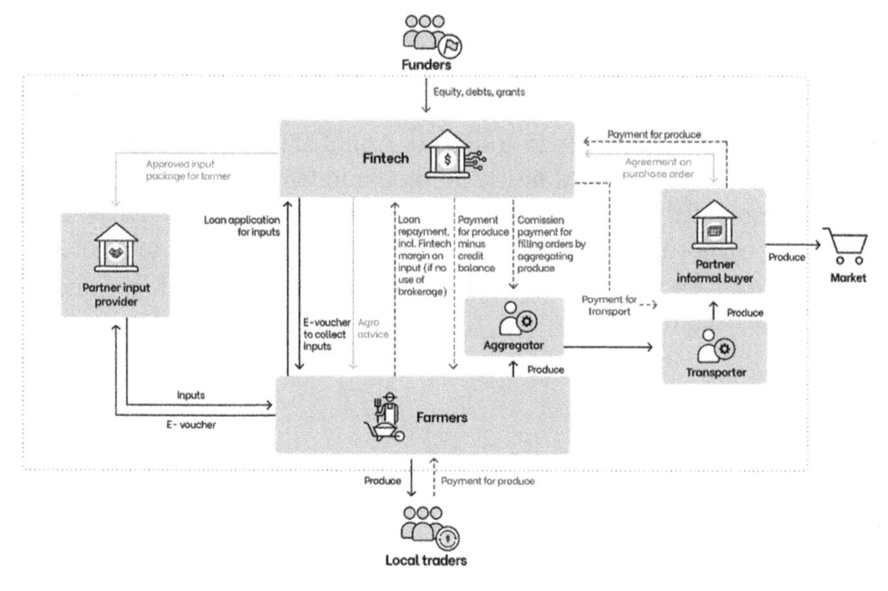

Fig. 3.1 Illustrative example of a financial service provider-led service delivery model with a holistic value proposition (*Source* Dalberg, Sustainable Trade Initiative [IDH] and MasterCard Foundation, Towards market transparency in smallholder finance: Early evidence from Sub-Saharan Africa, 2022)

only way to overcome the vicious cycle whereby credit is required to invest in a business's productivity, but increased productivity and income is needed to access credit. From a business model perspective, it seems that, in some circumstances, financial intermediary profitability could markedly increase if providers offer holistic value propositions.

Holistic value propositions can improve the economies of smallholder credit itself by enabling financial intermediaries to increase average loan size while managing risk. For example, from my work with IDH involved comparing financial intermediaries that added agricultural advisory services, access to inputs, and market linkages to those that did not. This revealed that, compared to those not offering additional services, financial intermediaries that did so saw, on average, farmer incomes ranging from 60 to 120 percent higher. This increased income can, in turn, boost demand for bigger loan sizes and reduce borrowers' risk. Additionally, as farmers increase their incomes and become consumers of more sophisticated financial services, providers can cross-sell nonagricultural loans (e.g., education or consumer loans).

By meeting a greater range of customer needs, holistic value propositions can strengthen financial intermediary differentiation, thereby nurturing customer loyalty. As the market for customers becomes more saturated and competition increases, borrowers' decisions may be driven more by accessibility and pricing than other factors. This is particularly true in highly digital ecosystems where credit is available at the touch of a button. Holistic value propositions enable access to a wider range of more tailored services, and this can be a powerful differentiator for customers. Finally, holistic value propositions can enable cross-selling of (potentially) higher-margin services to customers, such as insurance. However, bundles can also be a double-edged sword. Bundles that include products that clients don't fully understand, value, or wish to purchase, can negatively impact uptake and loyalty. More research is required to determine the right balance between fully bundled propositions and a menu of services.

A good example of a financial intermediary that offers holistic services to its customers is ECLOF Kenya. Through its CSA dairy loan program, ECLOF Kenya offers individual loans to farmers who are members of a partnering dairy collective and who have supplied milk for at least one year. The loan size ranges from $100 to a maximum of $4000, with loan terms ranging from 3 to 36 months, depending on the loan's size. The loan has a one-month grace period, and the payments are collected each month through a check-off system by the dairy cooperative at a flat interest rate of 1.5 percent per month (18 percent per year). It then combines the individual loan with nonfinancial services such as livestock insurance, veterinary support, climate-smart and agronomic training, and a guaranteed offtake from partner the dairy cooperatives. Research at Dalberg has shown that CSA dairy loan farmers performed up to three times better than regular dairy loan farmers who lack access to training or guaranteed offtake.[7]

Create pathways for customer growth

Too many financial intermediaries offer a standardized set of financial products that don't match the evolving needs of their customers. These standard products were designed based on a principle of risk mitigation and not with the intention of helping customers grow and achieve financial health. An alternative approach used by some financial intermediaries to help their clients is to design product pathways or staircases that adapt the size and type of financial services offered to clients as client needs evolve through growth.[8]

One such financial intermediary is Fonkoze, a microfinance institution based in Haiti. Fonkoze uses a "staircase out of poverty" model (see Fig. 3.2),[9] inspired by the success of a similar model offered by BRAC in Bangladesh,

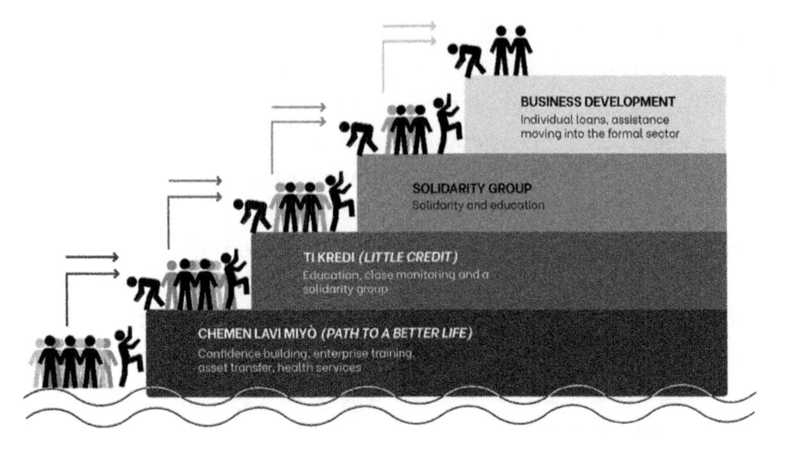

Fig. 3.2 Fonkoze's staircase out of poverty

that provides a comprehensive approach to poverty alleviation. The four steps or types of products and service offerings are uniquely designed to respond to the evolving and growing credit needs of women borrowers. The first rung of the staircase targets the ultra-poor, offering them the "pathway to a better life" service. Fonkoze provides these clients productive assets—such as livestock—along with materials to construct a basic home, access to free healthcare, and a cash stipend. In addition, each client is visited weekly to ensure they stay on track and use their time and resources productively. Once the client has developed a stable source of income, she moves up the staircase to the "little credit" product, the first loan with a repayment requirement. After a few cycles of these loans have been repaid, the women are encouraged to join a solidarity group that receives group loans ranging from \$50 to \$850. As they continue to grow their businesses, participants may take the last step to the "business development" loan, an individual 12-month loan starting at \$1300 and increasing from there as clients continue to succeed.

Along with the four main steps in the staircase, all clients also access "handrails": program support that includes business skills, general education, and access to health services for the women borrowers and their families. Fonkoze's model succeeds when a client graduates and no longer needs Fonkoze support.

Listen to customer voices to assess performance

As important as it is to design products and services that match clients' evolving needs, it is equally important for financial intermediaries to listen to customer voices when assessing their own performance. Listening to customers provides feedback to understand performance and customer satisfaction as well as input for feedback loops to improve financial service delivery moving forward. Financial intermediaries thus need to intentionally seek ways to collect customer views on their performance before, during, and after they provide them with services.

Collecting customer data should go beyond customer satisfaction metrics such as net promoter score (NPS), which are standard for larger financial intermediaries. It should also seek to understand how the finance is being used and whether it is adding material and social/psychological benefits to the customer's livelihood and well-being.

Here again Friendship Bridge provides a great example of how to put this into practice. For the last three years, Friendship Bridge has been working with 60 Decibels to listen to their clients' voices and collect client-level impact data about who their borrowers are and what benefits they perceive and experience as a result of receiving financing. Friendship Bridge has also been asking questions about customer satisfaction.

The results have been critical to the leadership team and board in identifying where Friendship Bridge creates impact and what strategic pivots it could make to amplify its impact. I share Friendship Bridge as an example because its social performance has been exemplary. By collecting this data, Friendship Bridge has learned that 35 percent of its borrowers live below the $5.50/day poverty line. They also learned that 94 percent of its customers earn more money and 48 percent consider their quality of life to have improved thanks to Friendship Bridge (see Fig. 3.3).[10] The next chapter looks more closely at this methodology and considers how other organizations might use it to boost their customers' financial health.

Measuring financial health

The above practices focused on how financial service providers better orient their services toward achieving better outcomes for their clients. However, I still haven't touched on the central issue of changing the metric of success around the provision of financial services. There is a need to move away from viewing the goal as access to financial services and toward building financial health.

Financial health is achieved when an individual's daily systems help build the financial resilience to weather shocks, and the ability to pursue financial goals.[11] It considers not only the access to basic financial services such

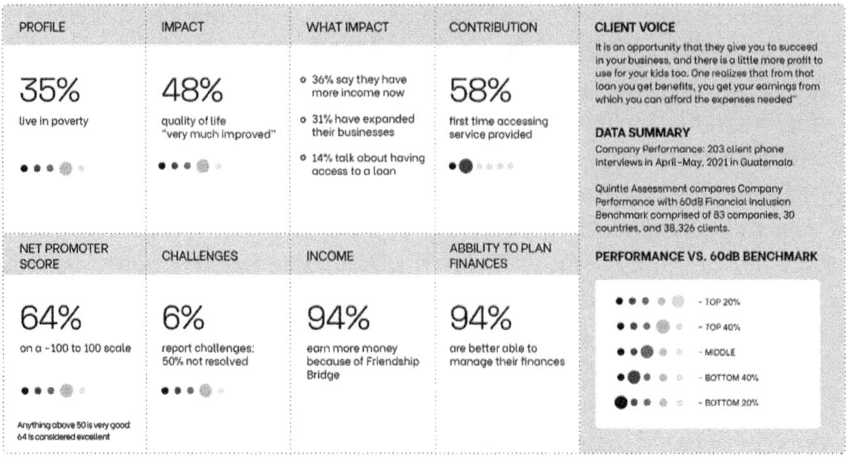

Fig. 3.3 Friendship Bridge social impact performance snapshot

as savings and loans, but also the behavioral aspects of consumers and their ability to manage their financial lives with dignity using a holistic set of services and how best to use them.[12]

Previous research by Dalberg helped to identify six dimensions that determine whether a consumer is financially healthy (Fig. 3.4).[13] The first is whether the individual can balance income and expenses. Financially healthy consumers are those that can shape their finances sufficiently to allow them to smooth consumption dynamically. Next, we must consider if the consumer is able to build and maintain reserves or savings. People save in many ways, both in cash and cash equivalents but also in assets. The ability to build those reserves correlates with wealth creation and ability to invest productively. Third, we should seek to understand whether consumers can manage existing debts and have access to potential resources. The ability to borrow from others which in essence assesses their social and financial credit can help consumers smooth incomes and make investments as well as influence how well or not they manage their existing debts.

Next, we should consider whether individuals are able to plan and prioritize financial goals purposefully. Assessing the ability and time horizon of an individual's plans goes a long way toward understanding if they have goals and ability to manage toward them. Fifth we should assess whether individuals can manage and recover from financial shocks. Apart from managing day-to-day needs, a financially health individual is prepared for the unexpected and thus we need to assess how well consumers respond to the now more frequent economics shocks that take place. Finally, we should consider whether consumers use an effective range of financial tools. Now

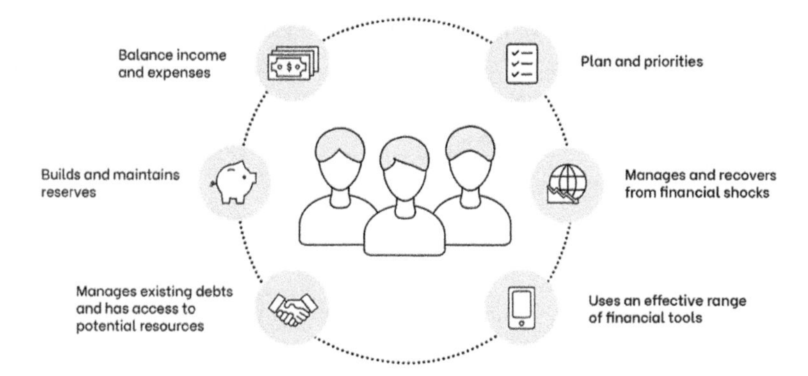

Fig. 3.4 Six dimensions of financial health

more than ever it is clear that savings and loans are not enough to lead healthy financial lives but rather additional services such as life coaching, insurance, and mobile payments play critical roles toward better and more resilient livelihoods.

Incorporating this broader spectrum of what success looks like for consumers, financial service providers will be able to create products and services that help build financial health. Let's now turn to learn how one measurement expert is doing this in practice.

Summary of key messages from this chapter

While no one-size-fits-all source of evidence describes what can lead to better financial health for financial intermediary customers, these five practices can help financial intermediaries do a better job and move in that direction:

- *Segmenting customers to understand user value* can help financial intermediaries target and tailor financial products and services to user needs.
- *Focusing on underserved customer segments* presents opportunity for growth for financial intermediaries and their borrowers.
- *Offering bundled or holistic services* that combine financial services with nonfinancial services tends to resolve more client challenges and improve client livelihoods while mitigating some of the key risks for the financial intermediary.
- *Supporting customers to grow* through a staircase model can help turn the poorest customers into the most successful.
- *Listening to customers to understand* performance and client outcomes is one important way to know if financial services are creating financial health.

To better measure financial health, financial intermediaries can also more intentionally move beyond access indicators to consider behavioral dimensions such as whether or not consumers balance their cash flows, build reserves, and plan for the future.

Notes

1. Dalberg, Sustainable Trade Initiative (IDH) and MasterCard Foundation, Towards Market Transparency in Smallholder Finance: Early Evidence from Sub-Saharan Africa, 2022. https://idh.foleon.com/farmfit/towards-market-transparency-in-smallholder-finance/. Accessed 29 Aug 2022.
2. Ibid.
3. VisionFund (2020) The Most Missing Middle Project Annual Report 2019–2020. https://www.visionfund.org/publications/report/most-missing-middle-project-annual-report-2019-2020. Accessed 11 Jul 2022.
4. Anne Maftei, Clara Colina, Understanding women's rural transitions and service needs, Rural and Agricultural Finance Learning Lab, 2020.
5. 60 Decibels (2021) Friendship Bridge Client Insights. https://www.friendshipbridge.org/wp-content/uploads/2021/08/60dB-@-Friendship-Bridge-Results-081021.pdf. Accessed 18 Jul 2022.
6. Ibid.
7. MacColl S, Riaz S, Colina C (2021) *Leveraging Risk Tolerant Capital for Loan Product Innovation: A Case Study of ECLOF Kenya's Partnership with Kiva*. Rural Agricultural Finance Learning Lab. Kiva Labs. https://www.raflearning.org/post/leveraging-risk-tolerant-capital-for-loan-product-innovation-case-study-eclof-kenya%E2%80%99s. Accessed 18 Jul 2022.
8. Ibid.
9. Fonkoze (2022) Fonkoze's Staircase Out of Poverty. https://fonkoze.org/staircase. Accessed 18 Jul 2022.
10. 60 Decibels (2021) Friendship Bridge Client Insights. https://www.friendshipbridge.org/wp-content/uploads/2021/08/60dB-@-Friendship-Bridge-Results-081021.pdf. Accessed 18 Jul 2022.
11. Ladha T (2017). *Beyond Financial Inclusion: Financial Health as a Global Framework*. Center for Financial Services Innovation.
12. Schoar A (2014). *The Financial Health Check A Behavioral Approach to Financial Coaching*. New America Foundation.
13. Ibid.

4

"Measure Impact with Client Voices"—An Interview with Sasha Dichter

This chapter summarizes my interview with Sasha Dichter, CEO and co-founder of 60 Decibels and a pioneer in promoting the practice of listening to client voices to understand impact.

Question 1: How did you first start to get involved in impact measurement?

Sasha: After a career spanning Booz Allen, IBM, and GE Money, I found myself drawn to social development work, and I found my way to Acumen Fund in 2007. I spent 12 years there, the first five as Director of Business Development, and the last seven as Chief Innovation Officer. While I was in the role of CIO, I increasingly focused on how Acumen could do a better job at collecting and reporting on its social impact. This led to the creation of Lean Data.

Question 2: What is lean data? What was the problem you were trying to solve?

Sasha: Impact has always been part of Acumen's DNA. In 2012, when I became Acumen's Chief Innovation Officer, one of the first things I wanted was to get under the hood of our social impact measurement and see if we could take it to the next level. And what I found was that despite our best impact intentions; despite having a high-risk, pro-poor portfolio; and despite having developed a piece of software, called PULSE, to collect social impact and operational and financial data from our investees; the core premise of

K. Hornberger, *Scaling Impact*,
https://doi.org/10.1007/978-3-031-22614-4_4

our approach was lacking. Our approach was to expect our investees to have and share their social impact data with us, but the fact was that, for us and for many impact investors, the actual impact data that our investees had was quite thin.

For example, although Acumen's mission is "to change the way the world tackles poverty," we did not have, at that time, objective, quantifiable data to understand to what extent Acumen was serving the clients that it was intending to serve—people living in poverty. In addition, we really didn't know from these clients whether they benefited from the products or services they received.

What we did know was a lot about the scale of these companies—the number of customers they served—or the number of jobs that they were creating. We knew this by relying, like many, on companies' operational data that we transformed into lives touched and potential impact created. For example, we might know, from an academic study, the typical impact of a solar home system, so we could use that to deduce that the solar home systems sold by Acumen investee likely had this level of impact.

While that's a good starting point, it is, at best, a very rough estimation of impact. And, if the study you're using is from a totally different geography or is out of date, your estimate of impact can be way off.

As a result, most impact "reporting" comes up short: it often serves as a self-affirming indicator of good efforts rather than an objective view of performance.

So, we decided to flip the idea from making data something we ask for and collect, to making a data offer to our companies.

We would collect data to help our social enterprises understand their clients better, and these data would tell us about their specific impact performance.

As mobile phone infrastructure was growing, we quickly landed on mobile phones to shorten the distance between us and the customer. We then built the expertise, methodology and the infrastructure to make it easy to do in practice.

The first thing we did was to collect data on poverty profiles of customers, and it worked! The moment we saw that data come back, a light went off that said, "this was the way forward."

So, in essence "lean data" is a philosophy around boiling down all the potential data you need to its essence, focusing on the data that you're most likely to use to drive action, and always thinking about the experience of the customer you're speaking to and making it positive for them. And, in practice, we found that the best way to do this was to use mobile phones to make it easy to speak to the people who matter most, listen to their voices and turn that into rich quantitative and qualitative data that enterprises can use. This allowed us to build an inherently scalable approach which could evolve into a data system and product for companies, and not just a handful of isolated studies.

Question 3: Why do you think it is important to do better impact measurement?

Sasha: Historically, impact was often treated as binary: Either it was occurring, or it was not. If an intervention was known to be "impactful"—often thanks to an academic study—it was typical to act as if the impact question had been settled, and, therefore, no additional data from the company was needed to confirm that it was, in fact, creating impact.[1]

This approach, though appealing and seemingly pragmatic, is fundamentally flawed. To understand why, think about the shift in our collective understanding of human learning and development that has happened thanks to the introduction of the concepts of "fixed" and "growth" mindsets. Those with a "fixed" mindset believe that intelligence is static. They prioritize looking smart and capable, and they therefore tend to avoid challenges. They are also more likely to give up easily when faced with obstacles and ignore useful negative feedback. Conversely, those with a "growth" mindset believe that intelligence can be developed, so they are more likely to embrace challenges, persist in the face of obstacles, and learn from criticism.

I see strong limitations of the "fixed" impact mindset every time we think we've discovered the "truth" about the impact of any given intervention (a microfinance loan, an improved school curriculum). It is, of course, helpful to uncover studies that show that an intervention works. But relying on studies alone and applying their findings to (somewhat) similar interventions ignores something essential: *differences in impact performance are the result of the specific actions of a company, the specific characteristics of a product or service, and the specific ways the company interacts with customers in a local context.* These differences, by definition, are invisible if one relies on a static definition of "this product or service (always) creates this much impact for a customer."

And yet, accepted practice is to tick the mental "yes" box on impact and then apply the "proven" impact variables to standard operational data. This can conjure up lofty impact numbers (such as number of people served, or lives touched), but the underlying, flawed premise of this all-too-common approach is the assumption that entire categories of interventions (e.g., every microfinance loan given to everyone everywhere) are essentially the same from an impact perspective and should be treated as such.

By contrast, an impact performance mindset is grounded in the knowledge that each intervention has a different impact in each place and each moment in time. Moreover, these differences are nearly always large enough to merit our attention. Understanding variances in performance, whether in impact or any other area, is the prerequisite for improvement.

Social interventions of all types exist in much too dynamic a context to blindly extrapolate from a single anecdote or study as broadly as we do. We would never assume that each supermarket, airline, or online marketplace has the same operational and financial ratios per unit sold. Yet we act as if the well-studied impact of one intervention in one place and one moment in time can represent all interventions of that type globally.

Question 4: Why is it important to listen to client voices when measuring performance?

Sasha: Current practice is to starve impact measurement of any real resources, thereby forcing practitioners to prioritize pragmatism above all else. Constrained in this way, we scramble to identify the data we can most readily produce and try to deduce potential impact from these data.

For example, consider the Global Impact Investing Network (GIIN) report titled, "Understanding Impact Performance: Financial Inclusion Investments." One would expect that such a report would be anchored around (1) the types of impacts that matter to a client of a microfinance or fintech organization; and that, (2) as a "performance report," it would allow one to understand the relative impact performance of financial inclusion impact investors and financial inclusion companies. Unfortunately, it accomplishes neither of these two objectives.[2]

How well does it map to the first objective: reporting on the sorts of impacts that matter to microfinance clients? Not very well. To see why, imagine the simple exercise of asking low-income microfinance clients why they borrow or save with a given MFI. These clients would likely tell you that they want to have more savings (so they can invest more), better withstand financial shocks, or be able to send their kids to school in the future; or they want increased income from growing a small business; or perhaps they are eager to have more say in family financial decisions.

Yet the metrics in the GIIN Financial Inclusion impact "performance" report do not include these sorts of simple, client-level impact metrics. In fact, of the 28 "impact" indicators in the report, only five of them bear even a slight relation to these essential client outcomes. This is because, as is typical, the report anchors in the data that are readily available—for example, number of loans disbursed or percentage of loans provided in local currency—and tries to deduce (potential) client-level outcomes from these data. But while the number of loans a microfinance organization has disbursed does give a sense of reach, it tells us nothing about the things we really care about: what's the impact of a microfinance loan on these clients of this MFI in this time period?

What we have in the GIIN report, then, is a very useful "state of the sector" snapshot that helps us understand the size of the sector, average loan sizes, portfolio at risk, and a bit of data about who is being served. This is all very informative. Almost none of it is "impact performance."

To understand, quantify, and, ultimately, improve impact measurement and performance, our objective should be to go directly to the stakeholders for whom financial intermediaries are meant to create better outcomes and, quite simply, ask them whether or not those outcomes are occurring. Even if this data is not perfect, even if self-reported data is often subjective, surely it is better to ask and get the data than it is not to ask at all.

Listening to client voices to assess performance must be anchored in the priorities of the affected stakeholders, have data gathered directly from those stakeholders, and allow for standardized comparison of the relative performance on the same set of objective metrics. This is a much better approach than trying to triangulate bad process data from multiple sources, which is the current standard practice.

Question 5: Why did you start 60 Decibels? What does it do?

Sasha: We started 60 Decibels to create a path forward on a scalable impact measurement approach that could give our sector real impact performance data, to drive improvement. Based on our experience with Lean Data at Acumen, we recognized we were onto something useful that many others beyond Acumen could benefit from. We had developed a group of paying clients who found our data really valuable, and we realized that we could best serve their needs and that of the broader sector if we spun Lean Data out into a new company. We completed this spin-out in 2019, and started a new company, called 60 Decibels.

We named the company 60 Decibels to signal that what we are focused on is to raise the customers' voice—60 Decibels is the volume of a typical human conversation. Everything we do is about listening to the voice of beneficiaries. If we are in the business of creating social change, we have to endeavor to make that change with people, not "to" people. This orientation requires us to listen, to really listen, to what matters to people and to whether and how our products and services are meeting their needs. At 60 Decibels, we try to make this as easy as possible.

We do this by creating a standardized, mobile-phone-based surveying methodology that is easy to implement and highly replicable across different contexts. We then built a global network of nearly 1000 enumerators in more than 70 countries so we could speak directly to clients, suppliers, and beneficiaries easily and in their local language, nearly anywhere in the world. Currently, our enumerator network can reach 5 billion of the 7 billion people in the world.

Today 60 Decibels is a global, tech-enabled impact measurement company that brings speed and repeatability to social impact measurement and customer insights. We provide genuine benchmarks of impact performance, enabling organizations to understand impact relative to peers and set performance targets. We have worked with more than 800 of the world's leading impact investors, companies, foundations, corporations, NGOs, and public sector organizations. 60 Decibels make it easy to listen to the people who matter most.

At 60 Decibels we believe it is high time we reject the premise that impact performance data doesn't exist when it does, in abundance. We can do better than the common, harmful deception that has gone on for too long in our industry: impact reports that do not report on impact, and performance reports that tell us nothing about performance.

Question 6: Would you agree that the financial health of customers is a more important goal for financial intermediaries than access to finance alone?

Sasha: Just as we believe that impact performance measurement must focus on outcomes and not outputs, we also believe that the goal of financial institutions serving underserved customers must go beyond providing access—the goal should be to enhance their customers' well-being.

Now, to be clear, no one creating a new fintech or microfinance organization is against this goal. Everyone wants to do more than just provide access. So, it's not a question of intentions, but rather a question of whether people regularly have access to the data to know whether or not they are enhancing customer well-being.

So, what we want to do is help these companies design products and services that are affordable and meet customer needs. It also means that the financial services, whether credit, savings, insurance, or payments, materially improves their lives from the customer point of view. In our work with financial intermediaries, we see lots of providers who really move the needle on customer well-being; and many others that really fall short. With the data in hand, it's easy to tell the difference between the two.

Question 7: What have you learned about the performance of financial intermediaries achieving financial health for customers by listening to their voices?

Sasha: Over the last two years through the COVID-19 pandemic we led a global effort to promote standard, comparable impact outcomes data for the microfinance industry. The results of that work were the recently released 60 Decibels Microfinance Index driven entirely by end customer voices. This groundbreaking initiative included aggregated data from the customers of 72 participating microfinance institutions (MFIs) across 41 countries, supported by 19 Founding Partners including Accion, Advans, BRAC, ECLOF, FMO, Fundación Netri, Global Partnerships, Grameen Foundation, Kiva, LeapFrog Investments, MCE Social Capital, Nordic Microfinance Initiative, Opportunity International, Pro Mujer, ResponsAbility, SPTF, Symbiotics, WaterEquity, and Women's World Banking. The project was financially supported by the Tipping Point Fund for Impact Investing and by Ceniarth.

The results of that research have led to many interesting insights, some that confirm existing expectations, and some that are more unexpected. Below are some that I think are most relevant:

- Microfinance is reaching people to create new access to financial services: More than half of the nearly 18,000 clients we spoke to were accessing a loan for the first time through the microfinance institution.
- One out of three clients report "very much improved" quality of life because of their microfinance loan, with this being equally true for men and women.
- Microfinance clients report higher than average capacity to deal with unexpected economic shocks. Two-thirds of the clients we interviewed stated they would not have difficulty covering an emergency expense equivalent to 1/20th of the growth national income per capita of the country they live in.
- Clients that reported business income increase also reported better household outcomes. This finding validates the core premise of microfinance: that clients can put loans to productive use in their businesses, and that business improvements will translate into improved household well-being.
- Clients using their loans for business purposes report relatively bigger improvements in financial management and resilience. Reporting better

performance in ability to manage finances, on levels of stress, not having challenge in making repayments and understanding the terms of their loan.

Question 8: What advice would you give to financial intermediary leadership teams hoping to maximize their impact and improve the financial health of their customers?

Sasha: This is such an important question. While I don't have a silver bullet solution on what works to improve livelihoods, I do think that we can all take a big step forward by getting our hands on the data that will, in a very direct, day-to-day way, help us understand if we're delivering real value to clients and improving their lives. I'd suggest the following three priorities:

1. Be careful in developing "impact performance" reports that don't allow you to assess performance and only report on output indicators such as number of clients or percentage of clients that meet a certain criterion.
2. Be transparent about how and where you collect data. In specific, seek ways to collect directly from the stakeholders impacted by your business, be they customers, employees, or the communities where you work. Listen directly to what they are saying and design your performance scorecards based on their voices.
3. Prioritize outcomes data such as whether income, business performance, or livelihoods materially improved because of access to financial services. While likely to be messy and hard to claim, attribution is incredibly important, if we seek to understand if financial products are working for customers or not.

Notes

1. Stanford Social Innovation Review (2021) This Is Not an Impact Performance Report, July. https://ssir.org/articles/entry/this_is_not_an_impact_performance_report
2. Ibid.

Part II

Provide Patient Capital, Not Venture Capital

5

Better Understand the Diverse Needs of Enterprises

The growth and limits of the Silicon Valley venture capital model

Over the last two decades, I have spent considerable time supporting entrepreneurs and aspiring entrepreneurs in honing their pitches asking investors to put capital into their ventures. Consider my friend Gabriel Migowski who in the late 2000s in Colombia had the audacity to start a new low-cost airline serving the Latin American market. In his case, given the large and growing addressable market, the ambition, connections, and growth mindset of the members of the founding team, they succeeded not only in raising the financing they needed but in making the business work. Their company, Viva Air, now holds almost 25 percent of the Colombian domestic commercial air market.

Viva Air is an excellent example of where venture capital can work and help businesses grow and succeed. At the seed stage, Gabriel and the three other founders secured $500,000 via ten tranches of $50,000 each from individual investors. Later, once they received regulatory approvals, they secured $30 million of Series A financing from a consortium of investors including Ryan Air, a Colombian conglomerate, and a Mexican transportation company. This then literally "fueled" their growth, enabling them to hire staff and to lease Airbus planes as their fleet. A decade later, they have successfully disrupted the airline industry in Colombia by making low-cost airfares not just something available in the United States and Europe but also across Latin America.

K. Hornberger, *Scaling Impact*,
https://doi.org/10.1007/978-3-031-22614-4_5

This is the logic of venture capital: Use early-stage private funding to back risky start-ups in the hope that big successes can carry a large portfolio that overall includes more failures than successes. Numerous examples illustrate this high-growth mindset, involving well-trained, well-connected teams who successfully use venture capital to grow their businesses in large and growing markets. Although VC-backed companies represent less than 0.5 percent of American companies created every year, they make up nearly 76 percent of the total public market capitalization of companies started since 1995.[1] This isn't just true in the United States anymore. Across the globe, venture capital is hitting record heights. According to Crunchbase historical deals data, global venture capital funding rounds have increased 12 percent per year in volume and 13 percent per year in value over the last decade (see Fig. 5.1). They have now mushroomed to more than 46,000 funding rounds totaling more than $800 billion in 2020 during COVID.

Venture capital misses far more than it hits: only one in ten investments (on average) provide a return.[2] But the sheer volume of transactions we see today means that a lot of 1-in-10 chances are coming up well. In fact, as of early 2022, more than one thousand privately held, up-and-coming companies were "unicorns" with valuations of $1 billion or more.[3] Of these, many are now in emerging markets such as Brazil (18), Indonesia (8), and Nigeria (3). Clearly, venture capital as pioneered in Silicon Valley is working for some.

However, the majority of entrepreneurs and aspiring entrepreneurs I have gotten to know and support have not been as successful with venture capital. This isn't always because they have a bad business idea, bad product-market fit, or bad timing. Rather their business model simply isn't and never will be

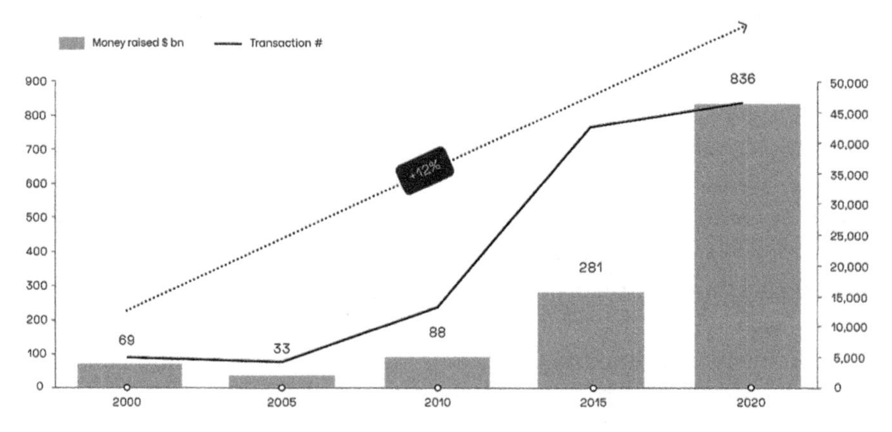

Fig. 5.1 Global venture capital activity by money raised and number of transactions, 2000–2020 (*Source* Crunchbase, Author analysis)

a good fit for venture capital. The risk-return profile of many start-up enterprises does not match with venture capital expectations. Traditional venture capital funds seek enterprises with hockey stick growth, large and growing total addressable markets, and founders typically from western elite universities with ambition to conquer those markets. This typically translates into expecting returns of between 25 and 35 percent per year over the lifetime of an investment.[4]

There are many impactful businesses that VCs reject because they don't fit the typical investment profile. For example, a charismatic Tanzania entrepreneur, Elia Timotheo started a fruit growing business in 2008, following his family heritage of farming in the Kilimanjaro region of Tanzania. He initially grew and sold apples and pears on a 150-hectare plot of land. Over the years, Elia recognized the problem wasn't a shortage of farmers but rather post-harvest loss, poor yields, and over-dependence on exports. Thus, in 2013, he founded a new B2B business called East Africa Fruits to improve the food distribution and to bolster the domestic market linkage for smallholder fruit and vegetable farmers across Tanzania. East Africa Fruits' innovation was to use cold-chain storage and novel mobile technology tools to better link buyers and sellers. Despite solving a clear market problem, Elia struggled to raise capital to grow his business for many years. Local angels and venture capital firms he approached repeatedly turned him down because they didn't believe in the market size potential, nor did they see Elia as a clear match to the traditional entrepreneur profile they prefer to back. It wasn't until 2017 when two impact investors—Fledge and Beyond Capital—got to know Elia and saw the potential of him and his business that he received his first seed investment for $125 thousand.[5] With that capital, he has grown the business, and while he still hasn't been able to attract traditional sources of funding, he has continued to raise funds from impact investors, with his most recent raise being $3.1 million dollars.[6]

Another example is an innovative but niche business concept I learned about in Colombia. Agruppa was founded by Carolina Medina and used mobile phone technology to empower small food vendors in low-income neighborhoods outside the capital by directly sourcing them with fresh fruits and vegetables at wholesale prices. The venture—which was born through a graduate school business plan competition and later supported through various accelerator and incubator programs, struggled mightily to raise financing to support the business. Throughout five years of effort, despite achieving sales of $1.2 million and securing some angel investments, the business never received venture capital and never took off. Eventually, Carolina gave up and the business closed. Perhaps her idea was before its time since it started in 2013, years before the COVID pandemic fueled the growth of the food delivery business platform and local sourcing models—but we will never know.

In fact, if we step back, we see that there is a much bigger problem. The world contains many more entrepreneurs like Carolina and Elia than like Gabriel. According to the World Bank's Consultative Group to Assist the Poor (CGAP), of the estimated more than 400 million micro-, small-, and medium-sized (MSME) enterprises, nearly 90 percent are microenterprises with between two and ten employees in low- and middle-income countries. These MSMEs contribute more than 80 percent of all jobs in those economies.[7] Worse yet, the International Finance Corporation (IFC) estimates that, globally, MSMEs face a $5 trillion financing gap annually, more than 1.3 times the current level of MSME financing available.[8]

The reality is that venture capital is not the right solution to meet a vast segment of this financing gap. To make their own business model work, venture capitalists' risk and return expectations don't match the underlying business logic and aspirations of the majority of MSMEs. At the same time, other traditional sources of financing, such as commercial bank debt, also fail to meet the needs of the majority. Thus, the financing gap persists, but must close to make the financial system work better.

To close the gap, a good place to start is by better understanding the diversity of enterprise needs through better market segmentation. MSMEs are diverse and so are their financing needs.

The Missing Middles

Within the MSME universe are a category of business I like to call small and growing businesses (SGBs).[9] SGBs have significant and positive impact on emerging and frontier markets. They create jobs, contribute to inclusive economic growth, provide underserved populations with access to essential goods and services, and spark innovative technologies and business models. These enterprises span a diverse range of sectors and business models—from rural agricultural cooperatives to innovation-driven start-ups to multigenerational small family businesses in sectors like retail, trading, and manufacturing—and are managed by an equally diverse range of entrepreneurs. SGBs typically seek external financing in the range of $20,000 to $2 million for various purposes—to support early-stage growth, expand operations, finance working capital, and acquire new assets—but they struggle to access forms of capital that meet their needs.

Accessing financing is particularly challenging for certain types of SGBs, such as early-stage ventures and businesses with moderate growth prospects that are stuck squarely in the "missing middle" of enterprise finance: They are too big for microfinance, too small or risky for traditional bank lending, and lack the growth, return, and exit potential sought by venture capitalists. Such businesses often face a fundamental mismatch between available financing and their specific needs. Addressing the SGB financing challenge is critical to promoting robust, broad-based economic growth, and to unlocking the potential of entrepreneurs to positively impact their customers, employees, and communities.

A primary cause of the SGB financing gap is that small and/or early-stage businesses are inherently hard to serve. Financial service providers[10] often have difficulty assessing the risk-return profile of enterprises in this space because the companies lack track records, have inconsistent or weak financial performance, and generally lack records about their operations and management. Even when risks are well understood, cost relative to investment return (e.g., high transaction costs and small ticket sizes) may prevent traditional financial service providers from seeing a strong business case for serving these segments of the market.

Another factor contributing to the financing gap is the lack of an effective, widely adopted segmentation approach that can be applied to a highly heterogeneous population of SGBs, allowing for better and more meaningful differentiation among enterprises and their financing needs. In the absence of such an approach, there is confusion in the market and misaligned expectations around risk, financial returns, exit prospects, and impact potential for SGBs. This, in turn, contributes to inefficiency in matching enterprises with the right sources of funding and financial instruments (on appropriate terms and at the right time, according to their business development stage).

A new way of segmenting enterprises to better understand their financial needs

With support from the Collaborative for Frontier Finance, Omidyar Network, and the Dutch Good Growth Fund, I developed an SGB segmentation framework that solves this challenge. The framework[11] offers a new way to understand the financing needs of SGBs in frontier and emerging markets. In contrast to previous segmentation efforts, which focused on a subsegment of SGBs (e.g., social enterprises serving low-income populations or enterprises in a particular market or sector), this framework covers the full universe of impact-oriented and traditional enterprises that have strong prospects for growth and job creation. Moreover, it segments enterprises using a mix of quantitative and qualitative characteristics that cut across both observable enterprise attributes and the behavioral traits of entrepreneurs.

Our research identified four relatively broad SGB "families" that occupy the missing middle, differentiated according to several variables that impact their financing needs as well as their attitudes to external finance. Each of these enterprise families tends to play a distinct role in driving inclusive economic growth and job creation in emerging and frontier economies. Each family also faces different gaps or mismatches in the market between available investment options and the solutions that are best suited to enterprise needs.

These four families come into focus when we look at the universe of SGBs through three distinct variables:

1. *Growth and scale potential:* An enterprise's prospects for future growth, their potential to reach significant scale, and their pace/trajectory of growth.
2. *Product/service innovation profile:* The degree to which an enterprise seeks to innovate in its core product or service offering or to disrupt the market in which it operates.
3. *Entrepreneur behavioral profile:* The entrepreneur's attitudes with respect to key dimensions that impact decisions on external finance, notably, risk tolerance, impact motivation, and growth ambition.

Four families of small and growing businesses[12]

The four families that form the top-level division of this segmentation framework are High-growth Ventures, Niche Ventures, Dynamic Enterprises, and Livelihood-sustaining Enterprises.

High-Growth Ventures

High-growth Ventures are SGBs that pursue disruptive business models and target large addressable markets. These enterprises have high growth and scale potential and tend to feature the strong leadership and talent needed to manage a scalable business that pioneers completely new products, services, and business models.

Often led by ambitious entrepreneurs with significant risk tolerance and a desire to achieve outsized impact, these firms begin as start-ups and, due to their rapid growth, soon "graduate" from SGB status to become larger firms. High-growth Ventures innovate by leveraging digital technology (e.g., social media platforms, mobile money transfer, etc.) but also by creating new hardware-based products and pursuing business model innovations (e.g., off-grid solar power, cookstoves, medical diagnostic equipment, etc.).[13]

Due to their steep growth trajectory, High-growth Ventures typically have a significant need for external financing. While SGBs make up a small percentage of an economy, High-growth Ventures have an outsized impact in driving innovation, spurring productivity, and creating new jobs.

Archetypes of High-growth Ventures include high-tech ventures (i.e., asset-light start-ups that are often software-based or digital and have favorable economies of scale); innovative businesses in established industries with "disruptive" potential (they may be tech-enabled but have a significant physical product or asset-intensive, brick-and-mortar component); and impact-focused companies that are pioneering new markets (e.g., serving the base of the pyramid) with the intent to achieve impact at scale.

An example of a High-growth Venture is Freight Tiger, a logistics tech company based in Mumbai that seeks to transform India's large transportation and freight industry through software that improves the end-to-end supply chain using end-to-end logistics visibility and cloud base tracking of freight movements. The company has secured multiple rounds of equity investment from top-tier venture capital firms in India.

Niche Ventures

Niche Ventures also create innovative products and services, but they target niche markets or customer segments. They seek to grow but often prioritize goals other than massive scale—such as solving a specific social or environmental problem, serving a specific customer segment or local community, or offering a product or service that is particularly unique or bespoke. Example archetypes of such businesses are creative economy enterprises that specifically focus on adding unique artistic value in niche markets and locally focused social enterprises dedicated to having a deep social impact at a local level.

Vega Coffee is a Niche Venture social enterprise in Colombia and Nicaragua that sells roasted coffee beans sourced from smallholder farmers directly to customers in the United States. A Kickstarter campaign helped the company achieve growth by providing initial seed funding to conduct pilot tests of its model. The size and saturation of the mail-order specialty or fair-trade segment within the coffee market will limit Vega's model, but the company is well-positioned to continue its steady growth and increase the incomes of its smallholder partners.

Dynamic Enterprises

Dynamic Enterprises, the third enterprise family, operate in established "bread and butter"[14] industries—such as trading, manufacturing, retail, and services—and deploy proven business models. Many are well-established and medium-sized, having steadily expanded over a number of years. They seek to grow by increasing market share, reaching new customers and markets, and making incremental innovations and efficiency improvements—but their rate of growth is typically moderate and tempered by the dynamics of mature, competitive industries. Multigenerational family businesses are a common archetype of this segment, and entrepreneurs' behaviors are often influenced by the family members' attitudes toward growth, risk, and innovation. Dynamic Enterprises are often the backbone of local economies and are important sources of jobs for low- and moderate-skilled workers.

Dynamic Enterprise archetypes include local manufacturers with a strong local presence but limited reach into larger markets, established agricultural cooperatives that have export contracts with international buyers, or family-run restaurants with multiple chain outlets across a local or regional market.

For example, La Laiterie du Berger is a Dynamic Enterprise that manufactures dairy products from fresh milk collected from over 800 farmers in northern Senegal. The company has seen steady growth since it started in 2005 and supports the growth of the local dairy sector by collecting milk locally and providing producers with access to high-quality cattle feed, technical assistance, and credit to about 6000 small dairy farmers.

Livelihood-Sustaining Enterprises

Finally, Livelihood-sustaining Enterprises are small businesses selling traditional products and services. These businesses may be either formal or ready to formalize. They tend to operate on a small scale to serve local markets or value chains, often in sectors such as retail and services, and deploy well-established business models. Such businesses often start out as "mom and pop" shops at microenterprise scale, but subsequently grow incrementally to hire additional employees. Enterprises in this family are particularly important for sustaining livelihoods in rural and vulnerable populations. Their needs for external finance are small in scale, but many Livelihood-sustaining Enterprises can benefit from financial products that enable them to manage working capital.

Multigenerational small family businesses are a common archetype of Livelihood-sustaining Enterprises. These businesses have steady but modest upward growth and a few core employees outside the immediate family. A second archetype is a microenterprise that grows to hire employees beyond immediate family members, moves from informality to increasing formality, and seeks out capital beyond the scale of what most microfinance institutions can provide.

An example is Prime Auto Care Garage, a small, woman-owned business based in Kigali, Rwanda, that provides motor vehicle repair services. It has been in operation since 2001, has 27 employees, and received a loan from Business Partners International for additional garage equipment to help it grow.

Differentiating the four families

To highlight the distinguishing characteristics of each of the four families of SGBs, one needs to apply three "lenses." Each of these lenses provides a unique way of comparing these distinct categories of SGBs and helps to highlight differences in their financing needs.

Lens 1: Product Innovation vs. Market Scale Matrix

A product-market matrix logic (see Fig. 5.2) differentiates families by the type of product or service the enterprise seeks to offer to a set of target customers. The x-axis is the "product innovation profile," described by the extent to which enterprises seek to be innovative in their core product and service offerings or to disrupt the markets in which they operate. Traditional businesses provide mainly existing products to existing customers. Innovative enterprises pursue incremental innovations to products and services and market extension strategies to reach adjacent customer segments. Disruptive enterprises seek to create or pioneer new markets that meet new, unmet customer needs. The y-axis plots growth and scale potential, which maps the extent to which an enterprise has the potential to scale beyond the size of an SGB and serve larger markets and customer bases.

Using this framework, High-growth Ventures and Niche Ventures both have a high focus on innovation, but High-growth Ventures are differentiated by their very high-level potential and ambition to serve large addressable markets. Dynamic Enterprises and Livelihood-sustaining Enterprises typically pursue more traditional business models in established industries and differ primarily in their size, complexity, ambition, and potential for growth.

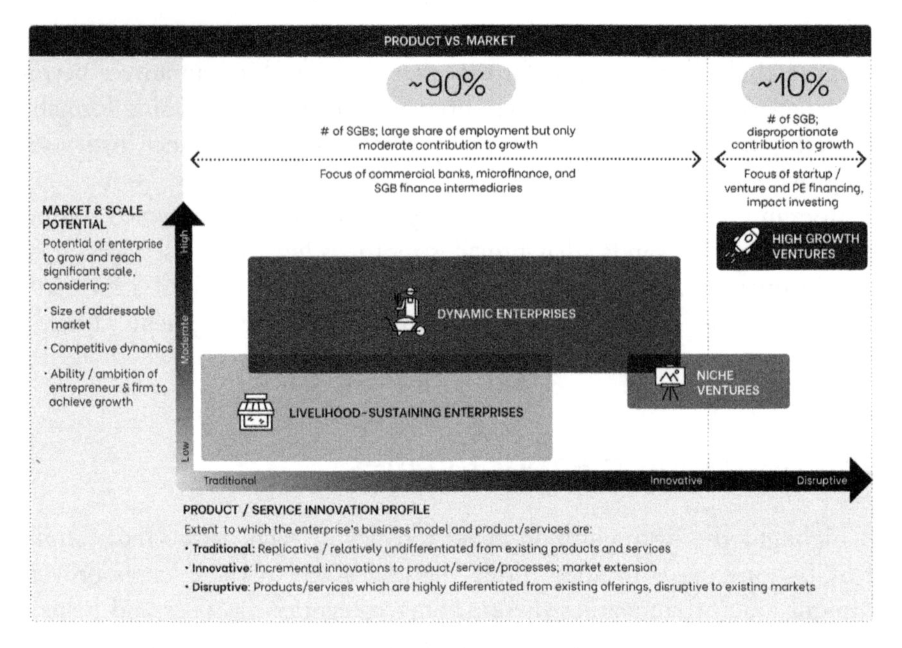

Fig. 5.2 Product innovation vs. market scale potential matrix

Lens 2: Growth Curves

Growth curves can be used to understand the paths of development and approximate size of the financing need of each of the four families over time. Drawing on quantitative portfolio data from five SGB investors, I used gross revenue and age of enterprise as proxies to see how the average enterprises in each family have evolved. The reason I opted for a simple depiction of broad growth trajectories over a set period (see Fig. 5.3) is that annual revenue growth is a highly contextual variable that depends on geography, sectors, and inflation levels, among other factors. These trajectories and time periods are illustrative and are primarily meant to show key differences in each segment family's respective growth trajectories.

The illustrative growth paths of each segment family over time reveal some key insights into their financial needs. A High-growth Venture, to be successful, must leverage early-stage risk capital and subsequent growth capital to scale beyond "SGB status" in a relatively short time, although the trajectory of growth for asset-light or tech businesses often differs from that of asset-intensive or physical product-based businesses. Niche Venture businesses enjoy early-stage growth but differ in not reaching an inflection point of scaling; thus, they typically do not need larger tranches of growth capital.

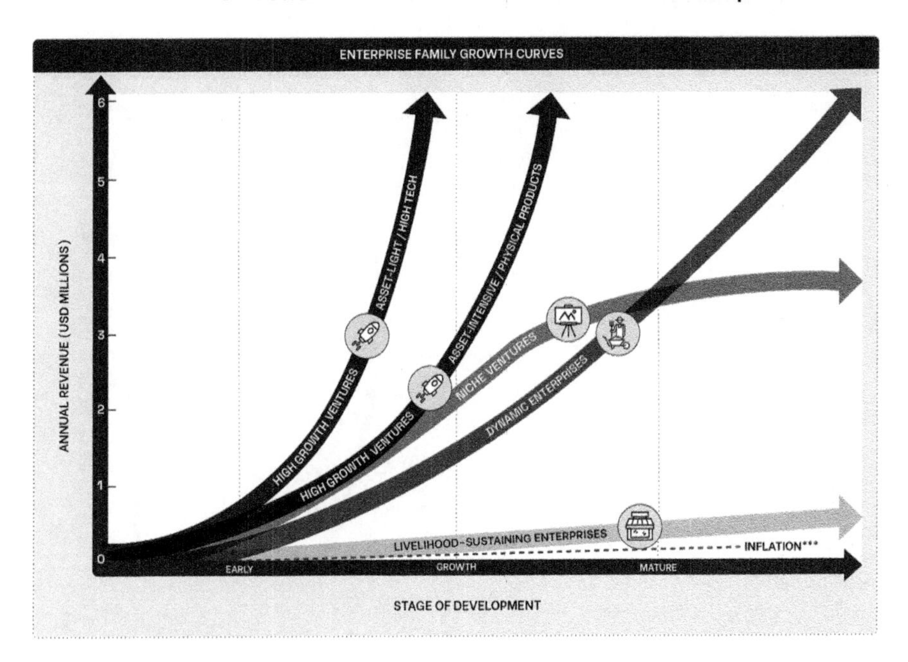

Fig. 5.3 Growth curves of four enterprise segments

Dynamic Enterprises tend to demonstrate more modest growth over a longer time horizon, while Livelihood-sustaining Enterprises start and remain small, and therefore require only limited, basic types of external finance.

Lens 3: Entrepreneur Behavioral Profiles

This lens considers the behavioral attributes of entrepreneurs, using human-centered design research to describe and differentiate the four families, and focuses on behavioral traits that significantly influence an entrepreneur's attitudes toward external financing—particularly attitudes toward risk, problem-solving motivation, and growth and scale ambition. Figure 5.4 illustrates "personas" for common management behavioral profiles in each of the enterprise families. These personas are not definitive, as entrepreneurs' behavioral attributes will of course vary from person to person and across demographic and cultural contexts. Rather, the personas are intended to illustrate attitudes common among entrepreneurs and management teams in these families as observed in the human-centered design research.

The figure describes High-growth entrepreneurs' persona as having high growth and scale ambition, the desire to problem-solve at scale, and the willingness to take risks to achieve that vision. The behavioral profile for a

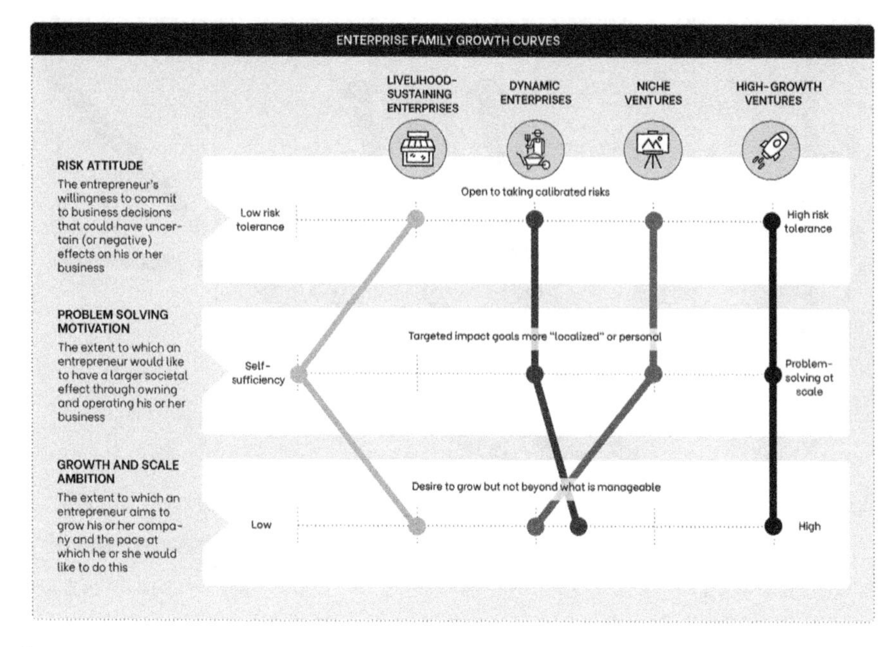

Fig. 5.4 Entrepreneurs' attitudes and behaviors toward risk, growth, and problem-solving

Niche Venture management team or founder represents the attitudinal factors central to what defines this family: the founder is willing to take risks (e.g., create a start-up with an "innovative" product offering) and seeks to achieve a specific vision of how to solve a problem or serve a particular customer segment. This founder has a specific vision for her or his business—achieving a specific kind of impact or designing a product that targets a specific niche market or customer segment—and prioritizes fidelity to that specific vision of success more highly than maximizing scale.

Leaders of Dynamic Enterprises are characterized by calibrated risk-taking to achieve success in the long term. Finally, for Livelihood-sustaining Enterprises, the business owner or management team typically has low risk tolerance and in decision-making gives considerable weight to factors such as stability and security—the responsibility of meeting family livelihood and immediate financial needs.

In conclusion, venture capital is not well suited to the needs of most small and growing enterprises across the globe. Rather, it is particularly well suited for High-growth Ventures—although not in all cases. The remaining three families of enterprises need the kinds of alternative financing approaches that will be discussed in the next chapter.

Summary of key messages from this chapter

- The venture capital industry has grown significantly and achieved some major successes over the last decade, yet its model does not match the needs of the majority of start-ups or enterprises.
- A better approach is patient capital that prioritizes impact, accepts risk, has a flexible time horizon for investment to ensure value is created; patient capital starts with understanding the diverse needs of enterprises.
- Small and growing enterprises typically fit into one of four broad families of enterprise: High-growth Ventures, Niche Ventures, Dynamic Enterprises, and Livelihood-sustaining Enterprises.
- Each of the four families is distinct in terms of its product innovation and total addressable market, growth potential, and the behavioral attributes of its founders and management teams.

Notes

1. The Economist (2022) The Bright New Age of Venture Capital. https://www.economist.com/finance-and-economics/2021/11/23/the-bright-new-age-of-venture-capital/21806438. Accessed 19 Jul 2022.
2. Andreessen Marc (2014) Big Breakthrough Ideas and Courageous Entrepreneurs. Interview Conducted by Stanford Graduate School of Business. https://www.youtube.com/watch?v=JYYsXzt1VDc. Accessed 19 Jul 2022.
3. Statista Research Department (2022) Global Unicorns – Statistics & Facts. https://www.statista.com/topics/5919/unicorns-worldwide/#dossierKeyfigures. Accessed 19 Jul 2022.
4. Zider Bob (1998) How Venture Capital Works. *Harvard Business Review*. https://hbr.org/1998/11/how-venture-capital-works. Accessed 19 Jul 2022.
5. East Africa Fruits. https://www.fledge.co/fledgling/east-africa-fruits/. Accessed 19 Jul 2022.
6. Dalal A (2019) Why We Invested: East Africa Fruits. FINCA Ventures. https://fincaventures.com/why-we-invested-east-africa-fruits/. Accessed 19 Jul 2022.
7. Kumaraswamy S (2021) *Micro and Small Enterprise (MSE) Finance: Examining the Impact Narrative*. CGAP. https://www.cgap.org/sites/default/files/publications/slidedeck/2021_03_SlideDeck_MSE_Finance.pdf. Accessed 19 Jul 2022.
8. SME Finance Forum (2019) *MSME Finance Gap*. IFC. https://www.smefinanceforum.org/data-sites/msme-finance-gap. Accessed 19 Jul 2022.

9. Aspen Network of Development Entrepreneurs. Why SGBs. https://www.andeglobal.org/why-sgbs/. Accessed 19 Jul 2022.
10. *Financial service provider* is an umbrella term for financial intermediaries and investors (local and international, traditional and impact-oriented) that directly invest in enterprises.
11. Dalberg (2018) *The Missing Middles: Segmenting Enterprises to Better Understand Their Financial Needs*. Collaborative for Frontier Finance. https://www.frontierfinance.org/missing-middles. Accessed 19 Jul 2022.
12. This section of the book is adapted from a report I previously wrote: Dalberg (2018) *The Missing Middles: Segmenting Enterprises to Better Understand Their Financial Needs*. Collaborative for Frontier Finance. https://www.frontierfinance.org/missing-middles Accessed 19 Jul 2022.
13. Koh H, Hegde N, Das C (2016) *Hardware Pioneers: Harnessing the Impact Potential of Technology Entrepreneurs*. FSG. https://www.fsg.org/resource/hardware-pioneers/. Accessed 19 Jul 2022.
14. "Bread and butter" in this context means essential industries that provide the basic necessities of life and form the backbone of business activity in developing markets.

6

Explore Alternative Approaches to Better Serve Enterprise Needs

A new paradigm of enterprise financing

The Silicon Valley model of venture capital established in the early 1970s focused on the quickly developing integrated circuits and computer hardware industries, and venture capital today continues to follow that traditional and rigid ten-year funding cycle. The underlying assumptions of this model no longer hold true, however, especially in emerging markets, leading to misalignment between companies and their investment partners. Real value is not created in the short term by generating the fastest return and focusing only on owners and investors; rather, it requires patience and consideration of the full range of stakeholders, shareholders, and employees and on the long-term interests and values of affected communities.

The current form of traditional venture capital follows commercial opportunity, rather than addressing the biggest problems faced by the world's most vulnerable people. Often laser focused on business opportunity and product-market fit, the model gives very little consideration to a business's underlying intentionality of effecting positive change. This approach will continue to lead to unequal outcomes for the majority and conservation of the status quo.

A better approach to venture financing for the majority of SGBs is something that the Acumen Fund terms *patient capital*.[1] Patient capital approaches venture investing by understanding and recognizing the diverse needs of the enterprise families discussed in the previous chapter. It prioritizes impact

K. Hornberger, *Scaling Impact*,
https://doi.org/10.1007/978-3-031-22614-4_6

over financial returns; it has a high risk tolerance, uncorrelated with financial rewards; and it maintains a more flexible time horizon for the return of capital than traditional 10-year closed-end funds.

The good news is that dozens of capital providers are now serving SGBs and taking a patient capital approach to providing financing. Over the last few years, I have gotten to know many capital providers that are changing the paradigm of how capital is provided. They are leveraging alternative financing approaches, such as altering the capital structure of the fund, using mezzanine instruments, and using improved technology to lower the cost of financing a broader range of growing enterprises. Together, these pioneers demonstrate how even the most persistent SGB finance gaps can be filled.

Five alternative approaches to closing financing gaps[2]

Through extensive research working with more than 50 alternative capital providers around the globe, I have identified five approaches capital providers can use to better serve High-growth and Niche Venture needs.[3] The approaches are differentiated by how they innovate on traditional finance models to meet the needs of both enterprises and investors buying into these enterprise categories. It is worth noting that the approaches outlined here can be used individually or in different combinations, depending on the needs of the enterprise the capital allocator is serving. The five approaches are summarized below; detailed descriptions follow later in the chapter.

1. **Open-ended capital vehicles**: This approach extends the time horizon using evergreen or open-ended legal forms, rather than traditional closed-end funds. It responds to the longer time horizons often required to create and capture value required for investments in agriculture, education, and health and other social services in emerging markets.
2. **Blended capital structures**: This approach blends capital that accepts disproportionate risk or concessionary returns to generate positive impact and to enable third-party investment that otherwise would not be possible.
3. **Customized financing products**: This approach adapts financial products to enterprise-specific needs and local market contexts. Highly customized "mezzanine" financial products are especially useful where straight equity or debt investments are difficult or do not match enterprise needs.
4. **Post-investment technical assistance**: This approach uses sector expertise and nonfinancial support to enhance the value of financing provided.

5. **Responsible exits**: This approach considers not just whether profit is maximized but also whether value creation/impact for all stakeholders is ensured on exit.

The product and operational model innovation characterizing each of these five approaches better meet enterprise needs by providing capital at the right time, of the right type, in the right amounts, on the right terms, and at an affordable cost to entrepreneurs. The needs of these enterprises vary by segment (as described in Chapter 5) and can be challenging for finance providers to meet for a variety of reasons, including the enterprises' differing risk-return profiles, misaligned expectations, higher costs of capital, longer return horizons, and lack of sophistication in local financial ecosystems. For example, the risk-free hurdle rate is so high in many emerging markets that it makes it really hard for small businesses to clear the hurdle.

Traditional providers, such as venture capital firms or commercial bank lenders, have business models that limit their ability to adequately and efficiently serve many enterprises' financing needs (as described in Chapter 5). The absence of innovative approaches to addressing those challenges leaves persistent core financing gaps. To a large extent, the five approaches listed above can augment traditional approaches and improve product-market fit through a combination of adjustments, including using different sources of capital and fund structures, deploying a wider set of financial products with greater flexibility, streamlining processes through standardization or digitization, or providing nonfinancial support.

To provide the right types of capital in the right amounts, many providers reduce risk and transaction costs by collaborating with local angel investors or by using products that balance customization and systematization. To provide capital on the right terms and at the right costs, some providers forge partnerships to provide value-added services, while others employ royalty-based products that tie cost to performance. The five alternative venture financing approaches I have identified highlight existing combinations of these innovations, but new approaches may employ them in novel combinations or combine them with traditional or as yet undeveloped components.

Open-ended capital vehicles

Recognition is growing that traditional closed-end fund structures don't match the time horizon of many enterprises' needs, particularly in emerging markets or with business models that confront systemic societal challenges. For example, in regenerative agriculture or forestry, it may take as long as seven to ten years for yields from the production of new crops or trees to mature responsibly. Education and health infrastructure can take years to find the necessary sources of revenue to build at the needed scale, as well as a product-market fit that justifies taking capital back out. Thus, the need is clear across the board for more flexible vehicles with longer investment time horizons for return of capital than most closed-end funds can currently offer.

Open-ended permanent capital vehicles (such as holding companies) allow asset managers to raise, deploy, and return capital on the time horizon that matches the realities of the ventures they are supporting, thus extending the capital deployment and return runway beyond 10 years.

Having the patience to wait for enterprises with impact to mature and achieve their potential is the cornerstone of patient capital. The pioneer of this approach is Acumen Fund, who uses their philanthropic backing to take a very founder-led approach to supporting their investees. They don't push their investees to liquidity until their investees are ready. For example, Acumen Fund is only now exiting Ziqitza Healthcare Ltd—which provides emergency ambulance services in India—after sticking with them for 18 years to see the business reach its potential.[4]

We see the open-ended or evergreen approach to fundraising becoming more common. From surveys we have run with capital allocators in Latin America, the Caribbean, and Sub-Saharan Africa, we find that the legal structure of roughly a fifth of the investors was an open-ended capital vehicle (Fig. 6.1). While most investors still value the simplicity and certainty of closed-end vehicles, this new flavor of asset management is compelling from an impact perspective and should continue.

A good example of this approach is Andean Cacao in Colombia. After years of struggling to secure traditional closed-end funding for a longer-term and perceived higher-risk investment to plant cacao in former war-torn regions

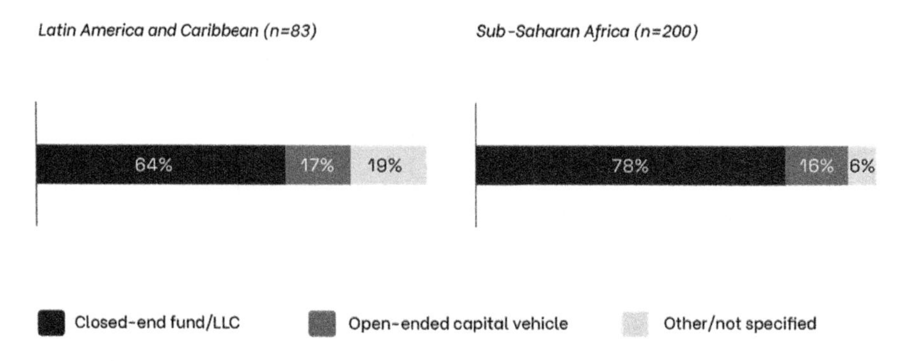

Fig. 6.1 Legal structure of investors surveyed, 2021 (*Source* Dalberg, ImpactAlpha, Author analysis)

of Colombia, my good friend Xavier Sagneires changed tactics. He created a permanent capital vehicle as a holding company via a joint venture with 12Tree and local farmers, now known as Andean Cacao. The venture raises capital from outside, securing capital from 12Tree and the IFC to plant and grow high-quality cocoa for export via sustainable partnerships with local communities. Through this model, Andean Cacao was able to start with an initial set of regenerative agriculture plantations and prove the model before adding funding that helped it expand operations. The key which made this model work, and will be required for others like it to succeed, is for the LP investors—in this case 12Tree and the IFC—to be comfortable with longer and open-ended structure and the implications it may place on their returns.

Sequoia provides another notable example of a venture capital firm recognizing the need to provide more patient long-term capital and support. At the end of 2021, Sequoia launched The Sequoia Capital Fund, an open-ended liquid portfolio made up of public positions in a selection of its companies. The new open-ended vehicle will in turn allocate capital to a series of closed-end subfunds for venture investments at every stage—from inception to IPO. Proceeds from those investments will flow back into the open-ended fund vehicle in a continuous cycle of renewable capital. In their own words: "Investments will no longer have 'expiration dates.' Our sole focus will be to grow value for our companies and limited partners over the long run."[5] The new structure removes all artificial time horizons and changes the way Sequoia partners with companies.

While open-ended vehicle structures do make a lot of sense for creating friendly relationships with patient entrepreneurs, they won't make sense in all circumstances such as when investors seek specific time-bound returns on their capital. When the simplicity and clear incentives that closed-end funds create are more appropriate, we can turn to other approaches to better serve a diverse set of short- and long-term High-growth venture needs.

Blended capital structures

A second approach to finding a better match for the financing needs of a more diverse set of enterprises is to blend the capital structure of the capital allocator. Blended structures allow asset allocators to better serve the riskiest but potentially highest-impact enterprises. These providers use grants or concessional funding to serve high-impact businesses, and since they are "impact-first," their returns to capital providers are intentionally concessional.

Blended capital structures typically involve at least two tranches of capital with different returns expectations, but they could have as many as five categories of investors. The blended vehicles often include senior or commercial investors seeking market rates or better returns. These are then catalyzed by one or more tranches of subordinated capital providers who seek anything from capital preservation to concessionary returns. The illustration below summarizes an example of a $50 million blended finance fund Dalberg designed in the health-technology space (Fig. 6.2).

A good example of a blended capital structure is the Impact-First Development Fund (IFDF) managed by Global Partnerships. Global Partnerships is an impact-first debt-fund manager based in Seattle. The IFDF used $5 million of subordinated first-loss debt and equity from an impact-first family office (Ceniarth LLC), foundation (W.K. Kellogg Foundation), and its own assets to secure a $50 million senior loan from a DFI (the United States International Development Finance Corporation, DFC).

Another, more complex, example is Novastar Ventures East Africa Fund I & II. The first fund focused $80 million on investing in early-stage businesses providing essential services to some of the poorest communities in East Africa. To make the fund come to life and make its intended investments, Novastar built a blended capital structure of investors. It had almost $15 million in junior or concessional equity from DFIs (the Netherlands

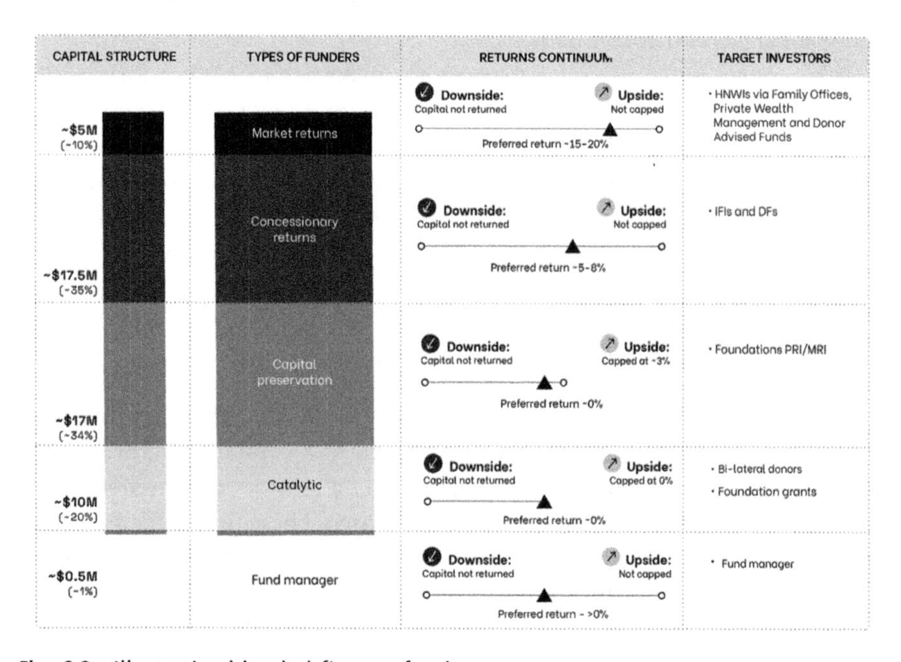

Fig. 6.2 Illustrative blended finance fund structure

Development Finance Company, FMO) and (Fonds D'Investissement et de Soutien aux Entreprises en Afrique, FISEA). Layered on top was $50 million of senior equity from other institutional investors (British International Investment, Dutch Good Growth Fund, European Investment Bank, and Norfund), and private investors (JP Morgan Chase & Co).

These two examples illustrate how blending the capital structure allowed capital allocators to focus on enterprise segments that other funders would not have been able to invest in due to lower risk appetite and fiduciary responsibilities.

Customized financing products

The third—perhaps the most complex, but very useful—approach to achieving a better match between venture investing and enterprise needs is by offering a customized set of financing products. Often referred to as "mezzanine" finance or "structured exit," this approach sits between the commonly deployed pure debt and pure equity solutions (see Fig. 6.3). These contracts take the best parts of equity (accompaniment, flexibility, partnerships, and patience) and mix them with the best parts of debt (self-liquidation and cost), without pressuring founders to achieve exponential growth and force the eventual sale of their business to satisfy investors.[6]

Mezzanine finance, described as "debt financing with equity-like features" or "equity financing with debt-like features," offers a suite of debt-equity hybrid products. On the debt with equity-like features side, it gives the lender the right to convert to an equity interest in the company in cases of default,

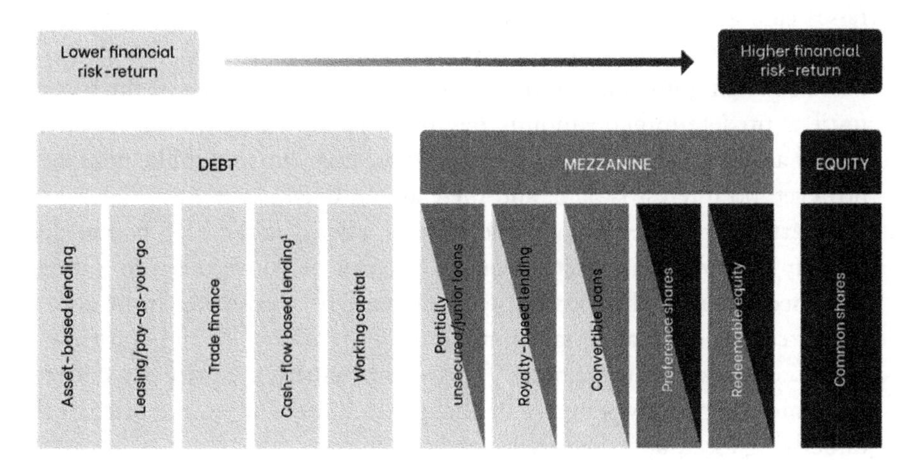

Fig. 6.3 Menu of different types of financing products

after more senior lenders are paid since mezzanine financing is senior to pure equity but subordinate to pure debt. As a result, mezzanine loans provide more generous returns than more senior debt, while also entailing more risk. On the equity with debt-like features side, it offers equity holders seniority or preference over other shareholders, which allows debt-like certainty of returns. In this way, investments are somewhat protected or typically less risky than pure equity investments.

Mezzanine financing is frequently associated with acquisitions and buyouts, where it can be used to prioritize new owners ahead of existing owners in case of bankruptcy. Mezzanine loans are most utilized for expanding established companies rather than as start-up or early-phase financing. More recently, mezzanine has become a popular approach deployed by first-time fund managers with less experience in traditional finance who want to be innovative and nimble in response to entrepreneur needs. Despite the rise in these two scenarios, I believe mezzanine financing is still not well understood or deployed. Below are definitions of some of the most common "mezzanine" products:

- **Subordinated or junior debt:** Subordinated or junior debt is most often an unsecured loan that ranks below other, more senior loans or securities with respect to claims on assets or earnings. In the case of borrower default, creditors who own subordinated debt will not be paid out until after senior bondholders/noteholders are paid in full. Subordinated debt may be secured by assets, but it is always second in line to senior debt holders in the case of default or underperformance of the borrower.
- **Revenue- or royalty-based lending:** Revenue- or royalty-based financing raises capital from investors who receive a percentage of the enterprise's ongoing gross revenues in exchange for the money invested. In a revenue-based financing investment, investors receive a regular share of the income until a predetermined amount has been paid. Typically, this predetermined amount is a multiple of the principal investment. Unlike pure debt, however, no interest is paid, and there are no fixed payments.
- **Convertible note:** A convertible note is a fixed-income corporate debt security that yields interest payments but can be converted into a predetermined number of common stock shares. The conversion from note to equity can be done at certain times during the note's life and is usually at the discretion of the noteholder. As a hybrid security, the price of a convertible note is especially sensitive to changes in interest rates, the price of the underlying asset, and the issuer's credit rating.

- **Simple agreements for future equity (SAFE)[7]:** A SAFE is a commonly used form of a convertible note. It is an agreement between an investor and an entrepreneur that provides rights to the investor for future equity in the company, similar to a warrant, except without determining a specific price per share at the time of the initial investment. The SAFE investor receives the future shares when a priced round of investment or a liquidity event occurs. SAFEs are intended to provide a simpler mechanism for start-ups to seek initial funding than convertible notes and reward start-up investors with discounts on future shares.
- **Preferred stock:** Preferred stock is a type of share that does not confer voting rights but is paid out before common stock in the event of liquidation. Preferred stockholders usually have no or limited voting rights regarding corporate governance. In the event of liquidation, preferred stockholders' claim on assets is greater than common stockholders but less than bondholders. As a result, preferred stock has characteristics of both bonds and common stock, which enhances its appeal to certain investors.
- **Redeemable shares:** Redeemable shares are shares owned by an individual or entity that are required to be redeemed for cash or for another such property at a stated time or following a specific event. Essentially, they are ownership shares with a built-in call option that will be exercised by the issuer at a predetermined point in the future.

An example of a capital allocator leveraging revenue-based lending is iungo capital, an East African-based fund. Iungo capital developed a tailored model that combines multiple innovations. It uses partnerships with local angel investor networks as both pipeline sourcing and risk mitigation tools; tailors primarily revenue-based loans to provide working capital without collateral; and has developed a self-sustaining nonprofit technical assistance arm to provide pre- and post-investment support (including for financial accounting of revenues) at a low cost. The revenue-based loans allow them to align their performance with the high cash turnover cycle performance of their mostly traditional business borrowers.[8]

An example of a capital allocator leveraging SAFE is Elemental Excelerator. Elemental provides a new nonprofit funding model for climate tech growth-stage companies and projects. They offer two types of SAFE products. One is a $500,000 SAFE note to fund new climate technologies or new climate projects that benefit Hawai'i or California's frontline communities. The second is a $300,000 SAFE note for global climate projects that will be deployed within 16 months and include elements of equity in access for the communities where they are working.[9]

These examples show that creativity is key to addressing complex problems, and the ability to adapt traditional funding models is paramount. These moves should be made cautiously, however, as mezzanine approaches can add an onerous complexity to deal making. One should always consider whether traditional grants, debt, or equity are easier instruments to use in meeting financing needs. Mezzanine financing should be undertaken not as an experiment but rather as a solution to enterprise financing problems.

Post-investment technical assistance

Investors can add very important value to enterprises they support by complementing capital investments with nonfinancial technical assistance. For traditional venture capital, this often involves sitting on and providing strategic advice to the company and/or CEO. While this is indeed valuable, some technical assistance can create even more value for a set of stakeholders that extends beyond just shareholders. Nonfinancial support can help professionalize operations, improve governance, introduce potential partners, actively broker potential exit opportunities, and strengthen enabling infrastructures. Notably, it can also include supporting enterprises to adopt stronger environmental, social, and governance (ESG) practices and policies.

Omnivore Venture Capital, for example, uses this approach to meet the financing needs of entrepreneurs building the future of agriculture and food systems in India. Started by two first-time fund managers, Omnivore's founders identified an underfinanced segment of early-stage enterprises disrupting upstream agri-tech and downstream agri-processing sectors in rural India, which traditional actors typically overlooked due to the challenging sector and geographic focus. Coming from nontraditional business backgrounds with significant sector-relevant expertise, these managers have not only been able to identify truly disruptive firms, but they have also been able to provide effective technical assistance on marketing and supply chain management to support the rapidly expanding firms in their portfolio companies.[10]

Pomona Impact provides another example. Pomona Impact is an impact investment fund manager and foundation based in Guatemala. Through their impact investment fund, Pomona provides innovative financing to better serve and support high-impact entrepreneurs in Central America. Recognizing the importance of not only helping high-potential enterprises become investment ready but also supporting them post-investment to adopt good ESG practices, the Pomona foundation has developed a set of accelerated technical assistance programs for green-tech, food, and renewable energy companies and, with support from Dalberg, designed ESG training for their portfolio companies. These technical assistance packages have dramatically improved the companies' performance and Pomona's impact in a hard-to-invest geography.[11]

Responsible exits

A final approach to enhancing the impact of venture financing is perhaps the most theoretical but also the most important. I call this approach *making responsible exits*. A responsible exit is the practice of not just maximizing the returns of an investment upon exit but also of ensuring that the enterprise is set up for sustained impact through operations that benefit *all* stakeholders, rather than only shareholders. In some cases, this may mean choosing to sell the business to an acquirer who doesn't offer the greatest economic value to shareholders, if that acquirer promises to prioritize the best long-term impact in terms of sustained innovation and on-the-ground benefits for the enterprise's customers and employees. Too often the decision to exit is done irresponsibly and driven by the terms promised to the capital providers, rather than in the interest of the enterprise or its stakeholders.

Achieving a responsible exit is not possible without planning for it. For example, if an investor doesn't intentionally screen entrepreneurs on whether they are impact-driven, how can the investor ensure the enterprise will remain impact-focused post-exit? To ensure a responsible exit, venture investors should consider its implications throughout the investment's lifecycle. Figure 6.4 highlights some key considerations and levers for investors.

Pre-investment, during sourcing and diligence, investors should ensure the entrepreneur's vision is aligned with the impact thesis, and that it will sustain, since the founder is a primary driver. To assess the risk of mission drift in

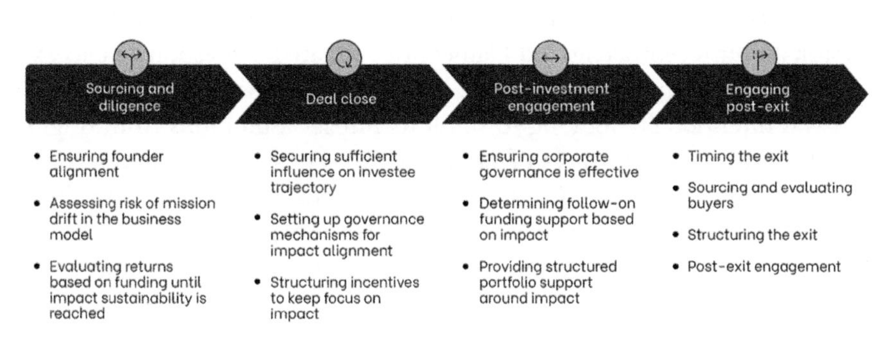

Fig. 6.4 Considerations for responsible exit by stage of the investment cycle

the business model, investors should consider to what extent the impact thesis is embedded in the enterprise and in what ways they can align market imperatives, which otherwise might shape the trajectory of the impact thesis. When evaluating returns, investors should focus on funding—until impact sustainability is reached.

When negotiating and closing the deal, investors should ensure a minimum threshold of ownership as a way to gain influence throughout the enterprise's trajectory. This is also an optimal stage for investors to set up governance requirements that align impact. For example, investors can require investees to comply with certain policies and report on certain metrics as part of their terms for investing. Another lever that investors have and should consider deploying here is designing incentives. Alongside traditional commercial incentives, investors can develop impact incentives to ensure continued impact alignment.

Post-investment, investors can ensure corporate governance and prevent harm in the long term by monitoring and maintaining governance standards. Investors can approach follow-on funding as leverage and motivate the investee to stay focused on impact. Early-stage companies are typically resource-strapped, offering early investors the opportunity to provide structured portfolio support around impact.

When deciding when to exit, investors should determine if the enterprise has reached impact sustainability. If the answer is no, investors should continue to provide impact support before pursuing an exit. With regard to sourcing and evaluating buyers, investors can cast a wide net and assess them based on impact alignment. Investors can also develop covenants in a sale to create penalties for drift from the impact thesis. Finally, post-exit, investors can continue to engage with investees and provide guidance on the impact thesis.

An example of an investor putting responsible exits into practice is the Omidyar Network in India (ONI). ONI has decided that making responsible exits is a critical part of living up to its commitment to be an impact investor, and it recently engaged Dalberg to evaluate how responsible its exits were from three venture investments. Dalberg found that, for the most part, ONI was intentionally looking to fulfill its impact aspirations from origination, demonstrating measures of performance against that aspiration, using its sectoral and ESG expertise to advise companies on how to adopt better practices post-investment, and working with company founders to identify buyers looking for something more than the greatest economic value.

Limitations

The five alternative approaches outlined above include examples of how asset managers can innovate to better support enterprises' financing needs. However, each of these innovations has limitations, and in many cases traditional approaches may still be preferable. Thus, by no means do I suggest that asset managers should adopt all five of these approaches. Rather, I argue that asset managers should start by understanding the financing needs of the enterprises they seek to support and then work backward to design the capital that best works to achieve the companies' goals.

While these five models primarily suit the needs of High-growth Ventures and Niche Ventures, and occasionally Dynamic Enterprises, other approaches better serve the financing needs of Livelihood-sustaining Enterprises and Dynamic Enterprises. These approaches include using internal knowledge and processes, or digital automation, to keep origination and servicing costs as low as possible are discussed later in the book.

Summary of key messages from this chapter

- Patient capital is an approach to enterprise financing that understands and recognizes the diverse needs of enterprises, prioritizes impact, has high risk tolerance uncorrelated with financial rewards, and maintains a more flexible time horizon to return capital than traditional closed-end venture or private equity funds offer.
- Five alternative financing approaches may better serve diverse enterprise needs, ranging from reconsidering the legal and capital structure of the fund to offering hybrid debt-equity financing products, enhancing investments with technical assistance, and considering a responsible exit pathway.
- Numerous examples demonstrate how these practices can be adopted.

Funders have important roles to play in helping to fill critical finance gaps by driving the adoption of alternative approaches and encouraging identification of new approaches and knowledge sharing.

Notes

1. Acumen. Patient Capital. https://acumen.org/about/patient-capital/. Accessed 19 Jul 2022.
2. This section of the book is taken, with some adaptations and with permission from the authors, from previous work published in the following report: Dalberg (2019) *Closing the Gaps—Finance Pathways for Serving the Missing Middles*. Collaborative for Frontier Finance. https://www.frontierfinance.org/closingthegaps. Accessed 19 Jul 2022.
3. Dalberg (2019) *Closing the Gaps—Finance Pathways for Serving the Missing Middles*. Collaborative for Frontier Finance. https://www.frontierfinance.org/closingthegaps. Accessed 19 Jul 2022.
4. Acumen Fund (2022) Investing as a Means: 20 Years of Patient Capital. https://acumen.org/patient-capital-report/. Accessed 21 Jul 2022.
5. Botha R (2021) The Sequoia Capital Fund: Patient Capital for Building Enduring Companies. https://www.sequoiacap.com/article/the-sequoia-fund-patient-capital-for-building-enduring-companies/. Accessed 20 Jul 2022.
6. Patton Power A (2021) *A Brief Guide to the Why, What and How of Structured Exits*. ImpactAlpha. https://impactalpha.com/a-brief-guide-to-the-why-what-and-how-of-structured-exits/. Accessed 21 Jul 2022.
7. This approach is described more fully at https://fundersclub.com/learn/safe-primer/safe-primer/safe/.
8. Learn more by visiting https://iungocapital.com/.
9. Learn more by visiting https://elementalexcelerator.com/equityandaccess/.
10. Learn more by visiting https://www.omnivore.vc/.
11. Learn more by visiting https://pomonaimpact.com/.

7

"Reimagine Approaches to Provide Capital"—An Interview with Chris Jurgens

This chapter summarizes my interview with Chris Jurgens, Senior Director at the Omidyar Network, a recognized thought leader on impact investing and reimagining capitalism.

Question 1: How did you first come to Omidyar Network and work on impact investing?

Chris: After spending more than a decade as a consultant and advisor with Accenture Development Partnerships and nearly five years at the U.S. Agency for International Development leading the office focused on public–private partnerships, entrepreneurship, and impact investing, I decided to seek new ways to impact the budding and growing field of impact investing. I'd long admired the pioneering work of Pierre and Pam Omidyar and the team of Omidyar Network in building the impact investing movement, and their approach to "investing across the returns continuum" to drive sector level change. And I was fortunate to join Omidyar Network in 2017. Omidyar Network was a prolific venture capital stage investor in impact enterprises in a range of sectors—including financial inclusion, education, and property rights—but they also had a wider commitment to field building. I joined the organization in a role focused on increasing the reach and effectiveness of the impact investment industry through catalytic investments, grants, and research. In that work I have had the incredible opportunity to work on efforts to mobilize catalytic capital to reach a wider range of small and growing businesses in emerging markets, and to work with great

K. Hornberger, *Scaling Impact*,
https://doi.org/10.1007/978-3-031-22614-4_7

partners including Collaborative for Frontier Finance and Catalytic Capital Consortium.

Question 2: Why do you think it is important to segment enterprise financing needs?

Chris: I think it's important to segment enterprise financing needs to address a persistent financing gap, and unleash the power of small and growing businesses to drive positive impact.

Small and growing businesses (SGBs) contribute to significant positive social and environmental impact in emerging and frontier markets: they create jobs, drive inclusive economic growth, spark innovative technologies and business models, and provide underserved populations access to essential goods and services. Yet SGBs in emerging markets face formidable challenges—most notably a lack of access to appropriate capital. Accessing financing is particularly challenging for certain types of enterprises, such as early-stage ventures and businesses with moderate growth prospects, that are stuck squarely in the "missing middle" of enterprise finance: they are too big for microfinance, too small or risky for traditional bank lending, and lack the growth, return, and exit potential sought by venture capitalists as we discovered through our joint research.

To move the needle on addressing the SGB financing gap, we at Omidyar Network believed the field needed to take a number of steps to better align the supply of financing with the demands and needs of entrepreneurs. And we thought a key starting point for this was to better understand and dissect this huge and diverse market. While a useful concept, the "missing middle" lumps together enterprises with vastly different business models, growth prospects, and financing needs, not to mention diverse entrepreneurs who have widely varying aspirations and attitudes towards external investment. Our hope is that better segmentation of SGBs will help investors, funders, intermediaries, and other actors in the SGB market to more efficiently identify financing solutions that address distinct needs of different types of enterprises.

By segmenting the SGB market into multiple "missing middles," we aim to more effectively diagnose the distinct financing needs and gaps faced by different types of enterprises, and in turn better focus on scaling the financing solutions that are most needed to empower enterprises of all types to meaningfully contribute to inclusive economic growth.

Question 3: What is Omidyar's approach to thinking about the continuum of capital in response to diverse enterprise needs?

Chris: At Omidyar Network, we have long believed that the impact investing market needs to deploy capital "across the returns continuum"—from fully commercial capital to catalytic capital and grants—in order to realize its full potential for impact. And Omidyar Network developed a specific framework for determining how and when we chose to invest in opportunities across different risk-return parameters in order to achieve our impact investing goals. Our investing experience showed that the relationship between returns and impact is complex and nuanced—and that investments in some sectors and solutions involve a trade-off between social return and financial impact, and some do not—depending on the conditions of the market, and the type of impact being sought.[1]

As a philanthropic investment firm, our core focus has always been to support organizations with the potential for significant social impact. In our early days, this led us to focus on supporting individual entrepreneurs and social innovators with the potential to scale solutions to deliver impact. But we quickly realized that individual enterprises—particularly those developing new solutions targeting challenging markets and hard-to-reach customer segments—faced a range of structural barriers to scaling. And that if we didn't take approach to develop entire markets and ecosystems around pioneering enterprises, that we wouldn't succeed. So, we increased our strategic focus on driving "sector level impact"—which meant not only investing in individual pioneering enterprises, but also taking a systems-level view of individual market sectors—like access to financial services in Sub-Saharan Africa—and identifying the key structural barriers to market development and scale. This then informed how we deployed investment capital across the returns continuum in a given context. We deploy venture capital investments on commercial terms in enterprises where markets are mature enough to support that scaling pathway. We deploy patient, catalytic investments where enterprises and markets are more nascent, and face significant scaling barriers, and may require a longer time horizon to reach commercial success and scale. And we deploy grants in areas like research, policy development, and field building where such approaches are critical to addressing the constraints that hold back an impact market from reaching its full potential.

Question 4: Which alternative financing approaches have you seen work best as alternatives to venture capital?

Chris: To address financing gaps I see promise in the growing number of financial service providers that are using alternative vehicle structures—such as open-ended or evergreen funds—that allow for more flexible time horizons. I am also encouraged by the growing use of alternative financing instruments—such as quasi-equity, revenue-based investing, and mezzanine debt—as alternatives to straight equity that allow for structured exits. Furthermore, we at Omidyar Network think models that mobilize local investors to provide early-stage risk capital—such as angel investing networks and seed funds that mobilize local LPs—offer potential to serve a wider set of SGBs in need of growth capital.

The research on "missing middles" that we conducted with Dalberg also highlighted a "transaction cost gap" that impacts the ability of Dynamic and Livelihood-Sustaining Enterprises to obtain affordable lending. The cost of providing low-ticket-size working capital financing or mezzanine debt can exceed the income that such products can generate for lenders. Here, we see tremendous promise in new technologies and tools being advanced by a growing cohort of financial services innovators that drive down the cost of credit assessment, underwriting, and servicing SGBs.

Nevertheless, it is clear there is no "silver bullet" financing solution to addressing the SGB financing gap. Rather, what is needed is a diverse, robust ecosystem of SGB finance providers that can meet the needs of different families of SGBs at different stages of their growth journeys.

Question 5: What about catalytic capital? What is it, and why is it important?

Chris: Catalytic capital is a subset of impact investing that addresses capital gaps left by mainstream capital, in pursuit of impact for people and planet that otherwise could not be achieved. It accepts disproportionate risk and/or concessionary return to generate positive impact and enable third-party investment that otherwise would not be possible. The capital gaps that it seeks to address are investment opportunities that mainstream commercial investment markets fail to reach, partially or fully, because they do not fit either the risk-return profile or other conventional investment norms and expectations required by commercial markets. I believe the "missing middles" are an excellent example of a set of capital gaps where catalytic capital can play an important role.

Question 6: What is Omidyar Network's philosophy behind responsible exits?

Chris: We firmly believe that impact investors must consider "responsible exits" as part of our investment strategy, in order to sustain impact beyond the time horizon of our investment. Taking on considerations for responsible exits is an integral element of intentionally building sustainable businesses that are set up to deliver impact.

Interestingly, our research in India shows that the levers for a responsible exit start well before the point of exit; investors have the greatest influence at the point of entry and during the term of their investment. A responsible exit must therefore include actions at each stage of investment to ensure that enterprises supported do no harm and that the impact generated can be sustained and amplified.

Question 7: What advice would you give to fund managers who are hoping to offer financial solutions to enterprise needs that are better suited than the traditional venture capital model?

Chris: I would encourage fund managers to focus on truly understanding the financing needs of the types of enterprises they are seeking to reach, and not just assuming that a traditional venture capital investment approach is the right product-market fit. This includes not just understanding their financial performance and the country or sector these enterprises operate in and their financial performance, but also the behavioral profiles of the founders as well as the types of capital that would best suit their business model. From there, key questions include: What types of capital solutions can best address those enterprise needs, based on the market conditions, the growth trajectory of the solution, and the goals of the entrepreneur? Are we best suited to offer that capital? How could we modify our approach to better serve capital gaps and create value for the enterprise and for our investors? This focus on enterprise rather than investor needs should help shift the narrative as well as encourage more asset managers to seek investors and investments that are better aligned to close capital gaps and create enduring impact.

Question 8: What are you focusing on now with respect to reimagining capitalism?

Chris: Building on our experience in impact investing and recognizing how traditional venture financing approaches are not able to solve critical capital gaps, we have recognized that there are larger systems issues we must confront. It is with the current form of capitalism itself. We believe capitalism can be a powerful force for good, but only when it is structured properly. Previous versions of American capitalism and some current versions of capitalism in Europe demonstrate that it is possible—and indeed optimal—to have a market-based system with freedom of opportunity that also allows for inclusion and basic economic security. Under an updated conception of capitalism, we must continue to incentivize and reward individual achievements. However, we must also ensure that people who have been historically and systematically marginalized will have opportunity, power, and the self-determination that comes from economic prosperity, in addition to a vibrant responsive democracy.[2]

With this objective in mind, we have rebuilt our focus on reimagining capitalism with five central pillars:

- Ground the economy in new ideas, shared values, and inalienable rights.
- Build an explicitly anti-racist and inclusive economy.
- Create counterweights to economic power.
- Rebalance the relationship between markets, government, and communities.
- Build a resilient economy that accounts for twenty-first-century context.

For the next few years, our shared task will be on these core areas and building a new capitalist system that will dramatically expand opportunity for the majority.

Question 8: What are you focusing on now with respect to reimagining capitalism?

Notes

1. Matt Bannick, Paula Goldman, Michael Kubzansky, Yasemin Saltuk (Winter 2017) Across the Returns Continuum. *Stanford Social Innovation Review*. Available online at: https://ssir.org/articles/entry/across_the_returns_continuum
2. Omidyar Network (2020) Our Call to Reimagine Capitalism in America. Available online at: https://omidyar.com/wp-content/uploads/2020/09/Guide-Design_V12_JTB05_interactive-1.pdf

Part III

Be an Impact-First Investor, Not an ESG Investor

8

The Difference Between ESG and Impact Investing and Why It Matters

Since the early 2010s, environmental, social, and governance (ESG) and impact investing strategies have grown significantly. On the ESG side, fund launches have increased 4× since 2012[1] which is more than twice as fast as non-ESG fund launches.[2] According to Bloomberg, by the end of 2022 ESG-related investments will reach $41 trillion globally.[3] On the impact investing side, trends analysis from the Global Impact Investing Network (GIIN) suggests that impact investing AUM has been growing by an annual growth rate of more than 18 percent and continues to accelerate year on year.[4] According to the IFC, impact investing totaled nearly $2.3 trillion in 2020 or about 2 percent of global AUM.[5]

Around the world, more people—particularly Millennials and Generation Zers—are seeking to deploy their money in ways that create financial returns while also responding to global challenges like poverty, disease, and rising inequality—issues that have been exacerbated by business-as-usual and passive investment decisions.

Now, thanks to the drive to achieve net-zero climate emission goals by 2050, retail investors in public markets are increasingly demanding ESG products. Annual growth of sustainable Exchange Traded Funds (ETFs) in public markets has been in double digits since 2019. A similar trend appears in private markets, where private venture capital investment in climate tech globally reached nearly $45 billion in 2021 up from just $6 billion in 2016.[6]

I have had a front row seat to the growth in the private markets. An increasing share of clients, whether development finance institutions (DFIs)

K. Hornberger, *Scaling Impact*,
https://doi.org/10.1007/978-3-031-22614-4_8

or institutional investors such as insurance companies, sovereign wealth funds, or pension funds, have been asking for advice on how to identify, grow, and diversify their ESG investments in emerging markets. This surge in ESG investing has been accelerated by investors' beliefs that ESG strengthens financial performance, improves brand image, and addresses societal imperatives while helping to manage long-term risks.[7] Additionally, a review of 1000 corporate ESG studies found that more than half saw a positive relationship between ESG and financial performance, and only 8 percent had a negative relationship.[8]

Yet, at best, ESG provides a weak signal of an investment's full environmental or social performance or risk; at worst, it allows greenwashing that diverts companies' attention from effective solutions to social and environmental challenges. Many companies with negative impacts but ESG facades are rewarded for good practices but don't deliver meaningful outcomes. According to one recent ESG rating provider, for example, the businesses of five of the top ten ESG-rated companies are in alcoholic beverages, industrials, metals and mining, oil and gas, and tobacco. Worse, in sharp contrast to the high correlation among credit rating agencies on their respective ratings, leading ESG rating providers show no or low correlation.[9]

Further, the ESG investing boom has yet to show it can contribute meaningfully to desired outcomes such as reduced carbon emissions. Global CO2 emissions continue to climb even as hundreds of ESG funds launch. While a time lag between deploying ESG strategies and reducing carbon emissions may be unavoidable, unreliable, and incomparable ESG ratings and insufficient auditing contribute to underperformance on these goals.

To make better informed choices and help recruit more capital into investments that will make a difference, we must start by unpacking the differences between ESG and impact investing, and why it matters. In many instances, these terms are grouped together, so by calling out the differences, we can create the room to help investors make better decisions on how to contribute to the world's biggest problems.

The ABCs of ESG versus Impact Investing

ESG investing is the incorporation of environmental, social, and governance (ESG) factors when evaluating an asset's performance. This type of investing involves a more holistic analysis than traditional investing and is often pursued as a means of improving investment performance. It focuses on the specific ESG factors most material to the asset's financial performance and evaluates how well the asset safeguards itself from risks posed by those ESG factors. For example, climate change threatens energy companies' bottom line, so energy companies are evaluated based on their operations' carbon intensity and efforts to manage climate-related risks. Thus, an energy company's reduced carbon footprint is driven by its desire to protect its bottom line, rather than solve the climate crisis for the benefit of the planet. Other common ESG measures include employee and management diversity, and whether or not a company has labor and worker safety standards in place. Independent rating agencies, such as Sustainalytics and MSCI ESG Ratings, measure companies' ESG performance using a scorecard. At the time of writing this book, there is no regulatory oversight in the United States around ESG ratings, so there is an important concern around their validity.

ESG is used commonly in public markets strategies and can involve screening out investments that do not meet ESG criteria. ESG criteria allow greater stakeholder advocacy because they force companies to track and report on how they are doing on a range of topics, including issues such as pollution and worker safety.

While there is overlap, impact investing is a distinct strategy from ESG investing. Impact investing strategies intentionally seek investments that contribute measurable solutions to the United Nations Sustainable Development Goals (SDGs). In this way, this strategy is focused on solving global challenges, rather than ESG's goal of protecting its bottom line. At present, impact investing strategies are commonly applied in private markets. In efforts to further define impact investing, the IFC developed the Operating Principles for Impact Management (OPIM), which are a framework of nine principles, adopted by over 160 investors globally. These principles include intent, contribution, measurement, responsible exits, and disclosure—all of which I discuss throughout this chapter and the previous section.

There is room for both ESG and impact investment strategies in the market, and investors may want to allocate funds to both in different proportions. Both can deliver superior financial performance and make the world a better place, but they work in different ways, although with some overlaps (see Fig. 8.1).

For example, most impact investors will screen for and manage ESG risk as part of their investment practices. Indeed, the OPIM requires this. By way of another example, ESG investing can generate the meaningful positive impact that impacts investing seeks to contribute. Specifically, corporate leaders may change their behavior because of ESG pressures, or ESG pressures may create

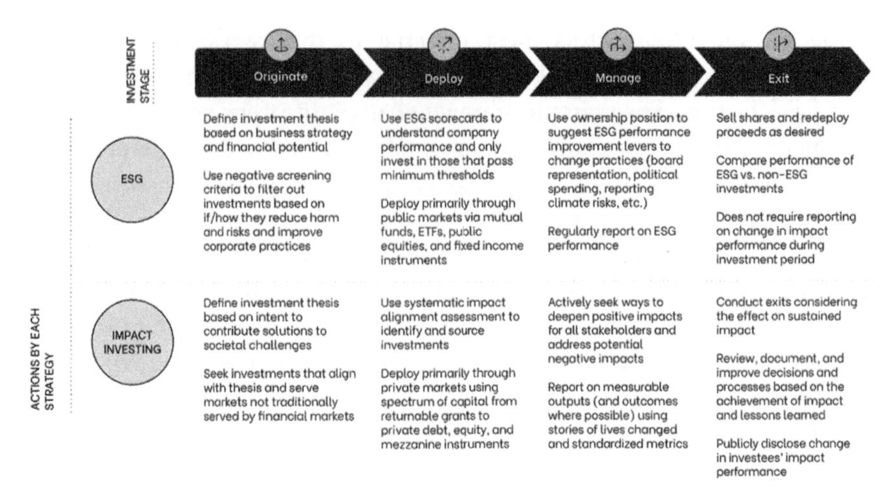

Fig. 8.1 Comparison of actions taken at each stage of the typical investment process by ESG and impact investing strategies

a big enough influence on capital markets to lower the cost of capital for firms generating positive impacts.

Despite these commonalities and the fact that many advisors lump ESG and impact strategies into a broader bucket of "sustainable" investments, investors can ensure their investments are impact investments and not ESG if they meet these three criteria: (1) assets selected with the intent to achieve impact; (2) the investment contributes to the impact of the investee firm; and (3) the impact is objectively measured.

Avoiding "impact washing"

These practices, enshrined in the OPIM,[10] help ensure that investments make a real, positive impact, and they help the market avoid "impact washing," which is when investors claim to have had an impact by taking credit for outcomes that would have happened anyway. Here is why:

Intent: Impact investors believe that capital can be purposefully invested to solve the world's most difficult and intractable problems, such as ending gender discrimination, combatting climate change, eliminating poverty, stopping systemic racism, and thwarting rising inequality. We won't solve these problems just by screening out companies in investment portfolios based on their potentially harmful practices. Rather, we must intentionally allocate capital to create solutions that directly solve these challenges. For example, BlackRock launched a $1 billion BlackRock Impact Opportunities (BIO) fund in 2022 to invest in companies and projects owned, led by, or serving Black, Latinx, and Native American communities, with the intent of solving global challenges around housing, financial inclusion, education, healthcare, inclusive transition, and digital connectivity.[11]

Contribution: Similarly, it is not enough for investors to hold shares in companies that manage ESG risk well or in companies that have a positive impact. Rather, investors' investments must make a meaningful difference to the companies' impact, either through financing their growth or by influencing their behavior. For example, Actis evaluates its contributions to each of its investments using a scale of low, moderate, or high based on its assessment of what the company's impact would have been in the absence of Actis's investment. In other words, Actis compares its involvement to the counterfactual to determine its true impact.

Measurement: Finally, we should not settle for good intentions and aspirations to contribute to the good. We should set realistic, evidence-based targets for what our investments can achieve, then monitor and evaluate actual achievements, using data and evidence to measure whether the investments we make cause material positive changes and avoid harm along the way. For example, Global Partnerships not only requires their investees to report on a set of common impact indicators, but they also run periodic in-depth impact studies, conduct field visits, and/or partner on external evaluations and assessments with firms such as 60 Decibels to more deeply understand and report on the impact their investments are having on the populations served.

ESG and impact on a spectrum

Another way to think about the difference between ESG and impact investing is to put them both on a spectrum of investment models that seek to achieve financial as well as environmental and social outcomes. Comparing ESG and impact investment strategies to traditional investors (see Fig. 8.2), one can see that ESG strategies continue to put financial considerations first but also evaluate, track, and mitigate environmental, social, and governance risks. Impact investment strategies on the other hand place more emphasis on environmental and social considerations by proactively seeking and managing assets for impact. Many larger asset managers, such as Apollo, BlackRock, KKR, and TPG, now have portfolios of funds that include both ESG thematic and dedicated impact funds.

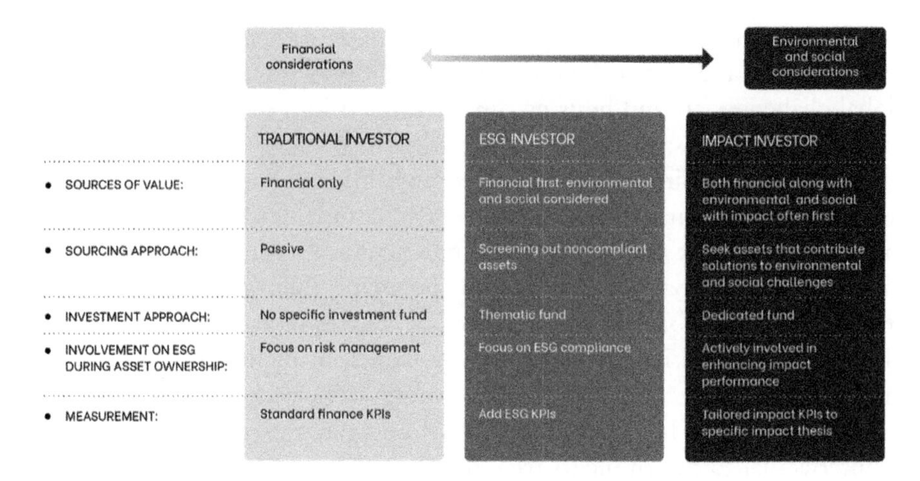

Fig. 8.2 Investors can pursue different investment models to achieve environmental, social, and governance goals

Achieving impact alpha

Adopting an intentional impact strategy is an increasing imperative for private equity growth funds. Rather than buy out and better manage investee targets, private equity is increasingly seeking value creation using other strategies. For many top-performing funds, the impact is an increasingly intentional part of the investment thesis and thus both reflects values and is actively managed as a driver of value creation. Many believe no trade-off is needed between positive social impact and financial return. Impact can create the alpha[12] that growth investors desire through various value creation levers previously not considered.

On the revenue side, private equity may seek an intentional impact strategy—such as selling carbon credits or using production waste for new products or sources of energy—as a way to consider a broader set of stakeholders and sources of revenue. It can also enhance investee branding, storytelling, and the investor's ability to connect with a new generation of customers, who are increasingly impact conscious. Providing details on how raw materials were sourced or how racially diverse employees are treated and supported can now be critical to marketing strategies to attract customers. Finally, impact strategies can help investees create value by attracting and retaining top-tier talent, a critical ingredient to any successful business.

An impact strategy can also help lower costs. It can lower the cost of capital for investees by helping them to attract new sources of funding with lower return requirements that may lower the overall cost of capital. Many funds are now co-investing with public or philanthropic sources of capital (more on that later), which inherently de-risks and lowers the cost of funding. The impact strategies can also lower operating costs by leveraging impact expertise to drive performance improvements within investees' operations. This doesn't necessarily involve cost-cutting but rather productivity-enhancing approaches such as using energy efficient technologies, data storage, sources of energy, etc.

Adopting these types of value creation approaches demonstrates that it is possible to both enhance financial performance and deepen/expand impact through capital allocation strategies. But how might this work in practice?

Putting it into practice

Now that we understand the differences between ESG and impact investing strategies, and how to avoid impact washing while generating alpha, it is important to understand how to put an impact investing strategy into practice. There are many useful frameworks out there, with the most notable being the OPIM (which I introduced earlier in this chapter) and Impact Management Platform (IMP), which was developed by IFC and other organizations and offer standards and guidance on sustainability management. There are also courses like Duke University professor Cathy Clark's Coursera on impact measurement and management for SDGs. However, none of these frameworks or courses really seem to provide enough details on how to do it in practice on your own. I thought it would be useful to share my approach, given that I've advised a diverse array of investors to develop their impact investing approaches.

I start by recognizing that financial returns and impact have the potential to be collinear (as demonstrated above), and thus I recommend that to develop an impact investing strategy, one should closely follow the steps one would take to deploy capital (see Fig. 8.3). My approach starts with defining strategic intent and then embedding those practices into your investment processes. Next, it moves to develop tools and resources to assess the expected impact at origination and how those tools and resources are used to monitor and enhance performance within investors. Finally, one evaluates the impact at exit and reinforces the impact by using feedback loops and independent verification to ensure the impact thesis and outcomes are continuously improving as capital is continuously deployed.

Let's now look at a step-by-step approach to putting this method into practice.

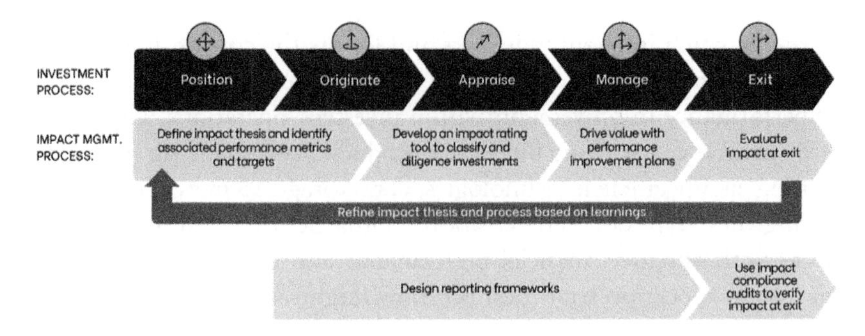

Fig. 8.3 Comparison of steps taken to deploy capital and manage impact

Step 1: Define impact thesis and identify associated performance metrics and targets

The first step in developing an impact investing strategy is to define the overall investment thesis for the portfolio. This should be a short and succinct statement that captures both the objectives of where capital will be targeted and how the capital will create financial value, once deployed. The investment thesis is not to be confused with investment criteria, which are specific terms and enterprise characteristics that the fund seeks when deploying capital and naturally follow from the investment thesis. An example of a clear investment thesis is from LeapFrog: "LeapFrog invests growth equity in high-growth, purpose-driven and emerging consumer-focused financial services and health-care businesses in fast growing markets in Asia and Africa, leveraging its unique hands-on value-add approach to support businesses in achieving rapid revenue growth, top-tier financial returns, and large-scale social impact."[13] In this thesis, LeapFrog clearly states its target investees and its intended value-add.

If you are focused on global development goals (which, I argue, is a good idea), the investment thesis should be mapped to the most relevant of the 17 United Nations Sustainable Development Goals (SDGs). Start by mapping how the fund's investments could contribute to different goals, prioritizing the SDGs that most closely align. For example, in LeapFrog's case, its fund might contribute most to SDG 1 No Poverty as well as SDG 3 Good Health and Well-Being, since its target investees provide a financial safety net as well as healthcare services.

Now comes the key effort: These SDG outcome aspirations must then be converted into a specific impact thesis that sits alongside and supports the investment thesis. The impact thesis is essentially a "theory of change" or a logic model that progresses from activities to outputs to outcomes, indicating how capital invested according to the investment thesis will contribute to the objectives of the SDG. A good example, sticking with LeapFrog, is its Financial Inclusion Fund II, which focused on SDG 1 No Poverty. The associated impact thesis was: "LeapFrog invests primarily in MFIs that deliver micro-insurance products (life, health, property, casualty), which provide a safety net to low-income individuals and businesses in case of external shocks and thus prevent them from dropping below the poverty line." LeapFrog clearly states how its activities (investment in MFIs) will contribute to outputs (MFIs have more resources to provide a safety net) and outcomes (MFIs' clients face less poverty).

As I always recommend to my clients, the impact thesis should clearly call out the five dimensions of impact, as stated in the IMP. These are What, Who, How much, Contribution, and Risk. Clarifying how these five dimensions apply to a proposed investment helps clarify who the target stakeholder is, what product or service they will receive, how much they will benefit, and the level of the investor contribution. It also helps clarify any potential risks that may arise. Figure 8.4 summarizes these five dimensions.

Once an impact thesis has been defined, it is imperative to identify and set quantitative targets for two or three associated key performance metrics (see the interview with Sasha Dichter to learn more about how to identify and measure performance). Note that I advise against setting just one metric since social impact should not be boiled down so conclusively. These metrics should be logically connected to the business activities of the potential portfolio companies, easily collected, and based on the strongest obtainable

IMPACT DIMENSION	IMPACT QUESTION EACH DIMENSION SEEKS TO ANSWER
WHAT	• What outcome(s) do business activities drive? • How important are these outcomes to the people (or planet) experiencing them?
WHO	• Who experiences the outcome? • How undeserved are the affected stakeholders in relation to the outcome?
HOW MUCH	• How much of the outcome occurs - across scale, depth, and duration?
CONTRIBUTION	• What is the enterprise's contribution to the outcome, accounting for what would have happened anyways?
RISK	• What is the risk to people and planet that impact does not occur as expected?

Fig. 8.4 Impact Management Project's five dimensions of impact

underlying evidence. The good news is that you don't have to search far for these metrics. For many years, impact investors have used different impact measurement standards. Now, because of the fantastic job done by GIIN and its team, the IRIS+ System has sufficient momentum to be adopted as the industry-wide standard for impact accounting. IRIS+ provides an industry-validated catalog of quantitative metrics (primarily output metrics) that can be used to account for the social and environmental performance of an individual investment as it pertains to an investor's impact thesis. To complement these quantitative performance metrics, users should also carefully think through the harder to quantify impact objectives and targets, such as strategic intent, and use IRIS+ metrics together with those objectives to think even more rigorously about trade-offs, transparency, and impact performance.

The last action necessary is to embed the impact thesis into the overall investment strategy. This will include efforts such as aligning fund governance with the impact thesis, perhaps by adding an impact chair to investment committees. Other useful considerations might include embedding impact logic into the fund's investment policy statement or investment guidelines, which lays out how the fund will ensure that impact considerations are incorporated into the entire investment process from origination to exit.

Step 2: Develop an impact rating tool to classify and diligence investments

The critical next step is to convert the impact thesis and key performance metrics into an impact rating tool to use in classifying and comparing expected investment opportunity impacts. These tools translate impact into a single common unit that allows users to identify impact-related trade-offs across investments, increases transparency, and forces a frank conversation about the assumptions used to estimate the impact of any single investment.

This is now often termed as developing an "impact due diligence" approach to parallel the commonly referenced "commercial due diligence" approach in private equity. Impact due diligence (IDD) should establish a set of impact criteria, an evaluation methodology, and a set of tools/questionnaires that can be used to collect information necessary to evaluate an enterprise for its potential alignment with the impact thesis. I intentionally use the word "alignment" because, unlike commercial due diligence, which seeks to assess whether the investment will *create* sufficient financial value, impact due diligence will rarely be able to sufficiently quantify the impact generated, and it is thus better to use it to assess whether the investment *aligns* with the thesis and potential for impact instead.

The first action is to define a set of impact criteria—ideally aligned with the five IMP dimensions—that will clearly lay out what the fund is looking for both quantitatively and qualitatively. For example, looking at the "who" dimension, an investor aligned with SDG 1 No Poverty, such as LeapFrog's Financial Inclusion Fund II, might consider whether the enterprise serves populations living in poverty. For these potential investees, LeapFrog would create an associated key performance indicator for that dimension, such as a number of client beneficiaries living below the international poverty line. This creates a clear and objective criterion for assessing potential investees on whether they are reaching people in poverty.

The next action is to develop a methodology to assess whether a potential investee aligns with the criteria, and if so, to what degree. There is no standard approach for how this is best done. I have had the opportunity to evaluate several impact measurement tools, including the IFC's Anticipated Impact Measurement and Monitoring (AIMM), TPG Rise Fund's Impact Multiple of Money (IMM), the Global Innovation Fund's Practical Impact, and the U.S. Development Finance Corporation Impact Quotient (IQ) tool. After carefully evaluating the pros and cons of each, I determined that the best tool presently available is the IMP impact classes, and believe it should be adopted as the industry standard moving forward. The IMP impact classes are also compatible with other impact rating and monetization approaches, making them the most flexible and usable for meeting a wide variety of investor needs (more on IMP in the Q&A with Olivia Prentice).

The IMP framework develops one set of questions and scoring criteria for each IMP dimension that can be scored using the objective criteria. These are merely guidance and ultimately it is up to the investor to define and refine their own methodology using IMP as guidance. An example of how IMP can be used as an impact rating tool is found in Fig. 8.5, based on a client engagement with MatterScale Ventures. However, for those interested in learning more, I recommend reading the Impact Due Diligence Guide from Pacific Community Ventures,[14] which goes much deeper on how this could work in practice or Section 2 of the Impact Frontiers Handbook which focuses specifically on impact ratings.[15]

Once the criteria and scoring methodology have been developed, it is necessary to develop resources for collecting data/information. Among these resources, I recommend requesting that the enterprise defines its impact narrative. I also suggest creating a simple and easy-to-complete impact data questionnaire for each enterprise, allowing you to collect enough information to assess the impact alignment of a potential investee without overburdening the enterprise.

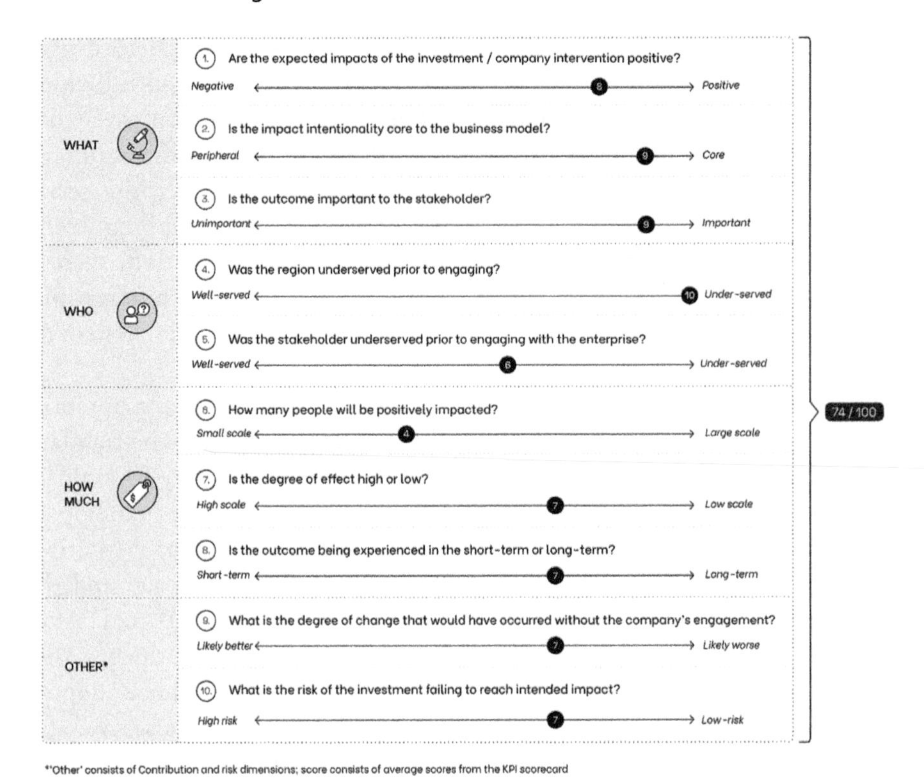

Fig. 8.5 Illustrative impact rating scoring tool at origination using IMP's five dimensions of impact

Step 3: Drive value through performance improvement plans

After the decision has been made to place capital into an enterprise that aligns with investment and impact criteria, the third step is to continuously monitor and enhance impact during the holding period. One way to do this is through an impact assessment framework, such as the B Impact Assessment,[16] which helps ensure and maximize the company's intended impact while also driving superior financial returns. These assessment tools inform development impact, supply chain, labor standards, and management or governance areas of strength and potential risks. Where risks are associated, corrective actions can be taken to mitigate them while ensuring impact outcomes are achieved.

Of course, for this to happen systems must be in place that can regularly collect relevant data about enterprise performance. This should be automated and as user-friendly for the enterprise reporting team as possible. Monitoring should be done on a semi-annual or quarterly basis and tracking what is happening with the impact performance targets and KPIs laid out in step 2. Most enterprises will need some support with data validation and interpretation, and in some cases, even data collection.

One consideration at this stage is that instead of treating impact performance and financial performance as two separate performance silos, it would be desirable to use metrics that make the material connection between the two explicit. Potential hybrid indicators like EBITDA/CO2 intensity for energy or EBITDA/yield per hectare for agriculture are two examples, but there are others. Hybrid metrics could help not only to improve impact reporting but also permit easier analysis of performance that rewards companies that perform best in both social and financial dimensions.[17] More on this in the next chapter.

Once the data are collected, the investor should compare performance with initial project impacts, any available benchmarks,[18] and whatever thresholds have been intuitively used to define the impact thesis. This initial analysis should also be complemented by a qualitative assessment of why the results are what they are. What happened during the holding period to date—which may or may not have impacted the investments performance—and what does it mean for the enterprise moving forward?

Finally, it is always important to systematically use the impact data collected when deciding what to do next. Hold the investment, or make a change to capital or impact targets? If areas of underperformance are identified, investors might consider ways to address the challenges if that can be done through technical assistance/expertise on turning underperformance around. If there are other factors at play it is worth assessing what this means for these types of investments moving forward.

Impact investors are also now learning how to monitor impact performance at the portfolio level, not just the individual deal level. To do this, one must normalize performance across investments, which may have very distinct impact on theses and KPIs. This is not an easy task, and more work needs to be done to develop streamlined approaches. However, once "normalized,"[19] investments can then be seen in concert with a portfolio, and thus patterns can be recognized and decisions about what to do next can be facilitated, considering not just the individual deal but the portfolio the fund holds.

Step 4: Design relevant impact monitoring and reporting frameworks

Once all the tools are in place, it is important to establish impact monitoring and reporting frameworks, processes, and tools. In efforts to not overburden enterprises, impact monitoring and reporting should be streamlined, done at a similar cadence to, and integrated with business and financial monitoring and reporting. Integrated metrics can also help organizations better align their core business activities and resource decisions, leading to more efficient and impactful investment decisions. In my opinion, the industry standard should be to update and integrate impact, financial, and operational metrics quarterly, with annual impact reports produced alongside or within financial performance statements. It is also critical that enterprise leadership see the value in collecting and reporting on impact performance. This can be assured by providing trainings that demonstrate not only how impact performance can be measured but also how it can be used.

A good reporting framework will start by using the investment organization's impact thesis and already defined performance metrics (ideally tied to the five dimensions of IMP) as the backbone for what it plans to report to investors and the public. The metrics should be easy to understand, prepared, and presented in a format that tells a story. I recommend showing performance over time and with case studies, as this is more interesting and revealing.

Ideally a good impact report will also integrate all relevant stakeholder perspectives on performance. Capturing not just the metrics/numbers that some LPs/investors may want to see but also complementing those with qualitative stories/insights from interviews with community members, employees, and customers. According to BlueMark—the leading impact auditor—less than half of the 31 reports evaluated included data or case studies that represented perspectives of stakeholders directly affected by investors' decisions.[20] Adding those perspectives is a critical ingredient to getting a holistic picture of performance.

Finally, impact reports should highlight risks, unintended impacts, and/or potential underperformance. Examples include shifts in consumer behavior, increased pollution, or changes in employee job satisfaction. These honest reflections are an important perspective on the true impact of both an organization and its investment capital. At present, I have yet to read an impact report that discussed negative impacts. This is not only dishonest but also undermines the credibility of the impact reporting systems. I urge readers who are involved with impact reporting to courageously lead the industry in disclosing unintended consequences. Learnings are an important teacher and a more holistic understanding shared by all will ultimately bolster the field, generating a more positive impact. Afterall good financial reports also mention business risks and how they are addressed, doing the same for impact risks seems like it should be good practice.

Step 5: Evaluate and reinforce impact at exit

The last step toward putting impact investing strategies into practice is also the last step of the investment cycle: exit and reporting of performance post-exit. For many investors, this is too often an afterthought. But it is mission critical to ensure impact is secured. Not only is it important to consider a responsible exit (see Chapter 6), but it is also important to use the learnings from the investment to improve the impact thesis and investment decisions moving forward.

It is also important to disclose externally how the fund has performed against its original intentions and target KPIs laid out in the first step. This builds transparency and credibility. An independent report should be prepared to share with investors at the very minimum, but ideally also with the public. The report should include annual deal and portfolio-level disclosure of all material impacts aligned with the investment thesis as well as indication of the sources of data used for reporting. Additionally, investors should require portfolio companies to report disaggregated indicators (e.g., female participation in management or race and ethnicity breakdown in leadership).

I suggest using impact compliance audits to verify impact at exit. This optional additional step entails hiring an independent third party to verify results and that correct practices are being deployed throughout the investment process. This process resembles how financial audit firms certify that financial statements comply with the common standard. I like the approach developed by Tideline, which assesses compliance, quality, and depth of each component of the OPIM or the UNDP impact assurance tool. The UNDP impact assurance tool is a scorecard that assesses impact considerations and results throughout the investment process, enabling better market comparison.

The top four accounting assurance companies—Deloitte, EY, KPMG, and PWC—all offer this service in addition to their auditing services, as do a range of boutique specialty firms, such as Dalberg, Tideline, and Vukani. Check out the verification summaries of the signatories of the OPIM for further information and ideas as well as whom to contact to pursue it.[21]

Is it working?

Now that we've clarified that impact investing constitutes a value creation investment strategy (which is unique when compared to ESG) and we've shown how it can be put into practice, it is worth reviewing whether impact investors are adhering to the operating principles and if the evidence on performance shows that it is achieving its lofty expectations.

On the question of adhering to the operating principles, it seems there is still considerable room for improvement. Private investors are only at the beginning of the professionalization of the practice of impact investing. In a review of 60 impact fund managers, only five (8 percent) were verified as adhering to all subdimensions of the nine IFC Operating Principles for Impact Management.[22] The vast majority—more than 85 percent—create logic models for an impact theory of change at the fund level as well as consistently review each investment's impact performance. However, less than 30 percent of funds reviewed assess each of the five IMP dimensions of impact, use composite impact rating methodologies to score investments, or solicit input from diverse stakeholders to assess performance.

To answer the question of whether lofty performance expectations are being met, we need to break the evidence down into two broad parts. First, we must determine whether existing impact investors are achieving expected financial performance, and second, we must determine whether they are achieving expected impact performance. Let's look at each in turn, with the caveat that the number of studies in the field remains limited, so conclusions discussed below may have shifted, depending on when you are reading this book.

In terms of financial performance, impact investing strategies are delivering on—but not exceeding—expectations. A recent Global Impact Investing Network (GIIN) review of financial performance found that nearly nine in ten impact investors in a survey of 89 funds that submitted data were meeting or exceeding financial return expectations.[23] Yet actual performance varies by asset class, market, and sector. The top performing is private equity in emerging markets. The average gross realized returns since inception for 34 funds making market-rate-seeking private equity impact investments in emerging markets was 18 percent, within a range of returns from 12 to 46 percent.[24] Another study by Cambridge Associates reviewed the financial performance of 98 private equity and VC funds that have a social impact objective and vintage years between 1998 and 2019. It found the net pooled internal rate of return (IRR) of those funds was 8.4 percent, although performance varied by vintage year, fund size, and market focus, with higher returns for post-recession years, funds with <$100 M assets under management, and investments focused on emerging markets.[25]

These are industry insider views based on self-reported data without comparison or counterfactual. If we look at the academic evidence, the story is more nuanced but still positive. A study by economists at the University of Chicago and Yale University showed that impact funds, when properly risk adjusted, underperform public market investment strategies but do no worse than comparable private markets strategies while also having substantially lower market beta than peer private funds.[26] Another study, from economists at the University of California at Berkeley, analyzed a sample of 159 VC and growth equity impact funds and found that impact funds have lower mean and median internal rates of return (IRR) than impact agnostic funds but that willingness to pay for those funds was strong, despite the lower returns.[27] Finally, the most positive and frequently cited study, from Harvard Business School in collaboration with the International Finance Corporation (IFC), showed that IFC's investments significantly outperformed equivalently timed public index funds focused on emerging markets (with as much as 30 percent higher returns).[28] It is important to note that this study only used IFC data, which may have played a role in driving that superior performance, given IFC's unique access to investment opportunities and lower market risk due to its association with the World Bank Group.

Looking specifically at private equity, I conclude that impact investment strategies can deliver market or above-market-rate returns. However, performance varies significantly (as it does for other investment strategies) based on a host of factors, including timing, market, sector, size of fund, fund management, and return expectation. It is also important to note that private fund investors seem to be willing to sacrifice financial returns for impact performance and, for the most part, seem satisfied with impact fund performance to date.

The news on the impact side I am afraid is less spectacular. The same GIIN survey mentioned above showed that all 89 funds surveyed reported they were in line with or outperforming impact expectations, yet they offer very little additional evidence to explain how that was achieved.[29] At present, most funds report and target output scale and reach indicators—such as lives affected or number of units sold—but from the outside, there are no good comparisons between funds on these metrics or ways to understand whether the numbers reported are good, great, or underachieving. We continue to rely on self-reported stories from the impact funds themselves. This will need to change if we are to truly realize the potential of impact investment. How to do this will be the subject of the next chapter.

Summary of key messages from this chapter

- ESG is an approach to investing that systematically incorporates environmental, social, and governance (ESG) factors material to performance. Impact investing is a strategy that intentionally seeks investments that contribute measurable solutions to global challenges.
- Three differences make impact investment preferable to ESG investment: impact investment incorporates clear positive intent, ensures clear logic for the contribution of investment capital to outcomes, and objectively measures and reports on impact.
- Investors can take five steps to put impact investing strategies into practice. This starts with defining an impact theory of change thesis and associated key performance metrics, followed by developing an impact rating tool to rate and classify investments against the impact objectives. Once capital is deployed, it is critical to manage for impact and to evaluate and reinforce impact goals at exit. Finally, it is important to create and report on impact performance as well as use impact compliance audits to verify performance.
- While evidence is nascent, research to date suggests that impact investment strategies are delivering expected financial performance in line with or superior to alternative private markets investment strategies. However, studies that benchmark and demonstrate impact performance remain limited.

Notes

1. Boffo R, Patalano R (2020) *ESG Investing: Practices, Progress and Challenges*. OECD. https://www.oecd.org/finance/ESG-Investing-Practices-Progress-Challenges.pdf. Accessed 28 Jul 2022.
2. Taylor T, Collins S (2022) Ingraining Sustainability in the Next Era of ESG Investing. *Deloitte Insights*. https://www2.deloitte.com/uk/en/insights/industry/financial-services/esg-investing-and-sustainability.html. Accessed 1 Aug 2022.
3. Bloomberg (2021) ESG May Surpass $41 Trillion Assets in 2022, But Not Without Challenges, Finds Bloomberg Intelligence. https://www.bloomberg.com/company/press/esg-may-surpass-41-trillion-assets-in-2022-but-not-without-challenges-finds-bloomberg-intelligence/. Accessed 28 Jul 2022.
4. Hand D, Dithrich H, Sunderji S et al (2020) 2020 *Annual Impact Investor Survey*. GIIN. https://thegiin.org/research/publication/impinv-survey-2020. Accessed 1 Aug 2022.

5. Volk A (2021) *Investing for Impact: The Global Impact Investing Market 2020*. IFC. https://www.ifc.org/wps/wcm/connect/publications_ext_content/ifc_external_publication_site/publications_listing_page/impact-investing-market-2020. Accessed 28 Jul 2022.
6. Pitchbook, Climate Tech Report, Emerging Tech Research, Q1 2022.
7. BNP Paribas (2019) ESG Global Survey 2019: investing with Purpose for Performance. https://cib.bnpparibas/esg-global-survey-2019-investing-with-purpose-for-performance/. Accessed 27 Jul 2022.
8. Whelan T, Atz U, Van Holt T et al (2021) *ESG and Financial Performance: Uncovering the Relationship by Aggregating Evidence from 1,000 Plus Studies Published Between 2015–2020*. NYU Stern Center for Sustainable Business. https://www.stern.nyu.edu/sites/default/files/assets/documents/ESG%20Paper%20Aug%202021.pdf. Accessed 27 Jul 2022.
9. Edmans A (2020) The Inconsistency of ESG Ratings: Implications for Investors. *Grow the Pie*. https://www.growthepie.net/the-inconsistency-of-esg-ratings/. Accessed 27 Jul 2022.
10. Learn more by visiting: https://www.impactprinciples.org/
11. Learn more by visiting: https://www.blackrock.com/us/individual/investment-ideas/alternative-investments/blackrock-impact-opportunities-fund
12. Alpha refers to excess returns earned on an investment above the benchmark return.
13. Learn more by visiting: https://leapfroginvest.com/our-portfolio/how-we-invest/
14. Pacific Community Ventures (2019) The Impact Due Diligence Guide. http://www.pacificcommunityventures.org/wp-content/uploads/sites/6/FINAL_PCV_ImpactDueDiligenceGuide_web.pdf. Accessed 27 Jul 2022.
15. Impact Frontiers (2020) Impact-Financial Integration: A Handbook for Investors. https://impactfrontiers.org/wp-content/uploads/2022/05/Impact-Frontiers-Impact-Financial-Integration-A-Handbook-for-Investors-Updated-July-14-2020.pdf. Accessed 29 Aug 2022.
16. See more by visiting the B Impact Assessment website: https://www.bcorporation.net/en-us/programs-and-tools/b-impact-assessment.
17. FSG, Hybrid Metrics: Connecting Shared Value to Shareholder Value: https://www.fsg.org/resource/hybrid-metrics-connecting-shared-value-shareholder-value/. Accessed 09 Oct 2022.
18. Some relevant benchmarks can be found at the World Benchmarking Alliance: https://www.worldbenchmarkingalliance.org/
19. Examples of common "normalized" indicators include lives touches, GHG emissions reduced or capital efficiency.
20. BlueMark (2022) Raising the Bar: Aligning on the Key Elements of Impact Performance Reporting. https://bluemarktideline.com/raising-the-bar/. Accessed 27 Jul 2022.
21. Learn more by visiting: https://www.impactprinciples.org/signatories-reporting

22. BlueMark (2022) Making the Mark: Spotlighting Leadership in Impact Management. https://bluemarktideline.com/making-the-mark-2022/. Accessed 27 Jul 2022.
23. GIIN (2021) Impact Investing Decision-making: Insights on Financial Performance. https://thegiin.org/research/publication/impact-investing-decision-making-insights-on-financial-performance. Accessed 27 Jul 2022.
24. Ibid.
25. Cambridge Associates (2021) Private Equity and Venture Capital Impact Investing: Index and Benchmark Statistics. https://www.cambridgeassociates.com/wp-content/uploads/2022/02/PEVC-Impact-Investing-Benchmark-Statistics-2021-Q1.pdf. Accessed 27 Jul 2022.
26. Jeffers J, Lyu, T, Posenau K (2021) The Risk and Return of Impact Investing Funds. Proceeds of Paris December 2021 Finance Meeting EUROFIDAI—ESSEC. http://dx.doi.org/10.2139/ssrn.3949530
27. Barber B, Morse A, Yasuda A (2019) Impact Investing. http://dx.doi.org/10.2139/ssrn.2705556. Accessed 27 Jul 2022.
28. Cole S, Melecky M, Olders F et al. (2020) *Long-Run Returns to Impact Investing in Emerging and Developing Economies* (Policy Research Working Paper, No. 9366). World Bank. http://hdl.handle.net/10986/34383. Accessed 27 Jul 2022.
29. SSIR (2017) Bono Doesn't Know—All of Us Are Still Learning. https://ssir.org/articles/entry/bono_doesnt_knowand_neither_do_the_rest_of_us#. Accessed 29 Aug 2022.

9

Raising the Bar on Impact Management and Measurement

The main criticism of both ESG and impact investing is that it is not easy to measure and compare investments using existing nonfinancial measures. Take, for example, a recent email I received from a leading finance professional at one of the world's largest asset managers. In the email she asked, with a hint of criticism, "How can I justify adopting an impact investing strategy if I have to choose among an alphabet soup of reporting standards, none of which clearly and consistently demonstrate results superior than if I was not to adopt them?".

This question indicates a broader problem that needs to be solved before impact investing achieves its potential. Unless we raise the bar on impact management and measurement, impact investing will always be a bespoke and niche strategy. A recent INSEAD survey of Limited Partners (LPs) stated that only 45 percent of general partner (GP) investees report any ESG or impact metrics and that most of them focus only on positive stories. The survey also found that less than one in six GP report granular impact data.[1] Another recent report, this one from BlueMark, reviewed 31 impact reports from existing impact investors and showed that fewer than half used standardized indicators, cited any sources, or defined metrics, limiting the comparability of reported data.[2] Despite this poor track record, the consensus is growing that private equity funds could and should take the lead on first standardizing ESG and impact accounting and then embedding those standards and metrics into their daily operations.[3]

K. Hornberger, *Scaling Impact*,
https://doi.org/10.1007/978-3-031-22614-4_9

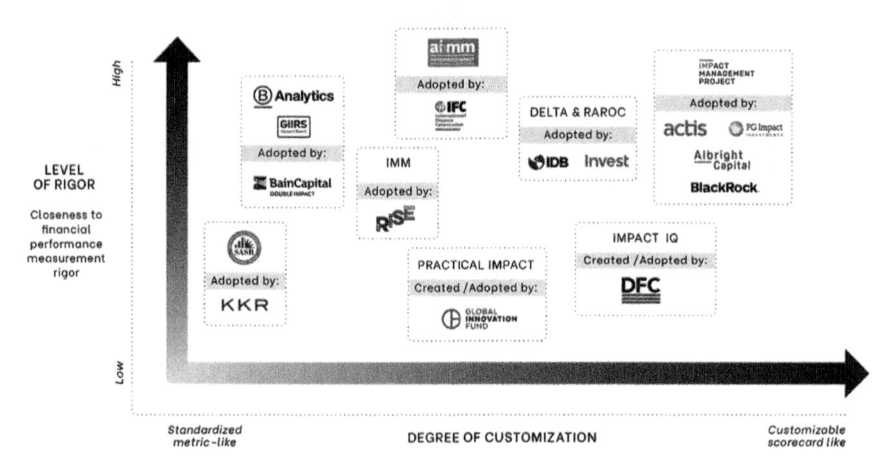

Fig. 9.1 Comparison of standard impact assessment methodologies currently in practice

The lack of common practices and standards means that at present impact performance reports and methodologies are primarily used to support fundraising and marketing efforts rather than for decision-making or to drive value creation. Therefore, most impact asset managers who are considering adopting or improving approaches have little incentive to do so. The only organizations changing the paradigm are those being pushed to do so by their boards or shareholders. This is a trend that will continue.

Yet once asset managers commit to adopting impact investing as a strategy, they confront a variety of impact measurement methodologies and inconsistent LP demands. At the time of writing this book, a variety of methodologies are used by private equity and development finance institutions, varying in terms of the degree of rigor and level of customization investors are allowed to deploy (see Fig. 9.1).

Three of the most well-regarded and adopted are the Impact Management Project (IMP) approach (now part of Impact Frontiers, and is the methodology I recommend),[4] the Impact Money Multiple (IMM) used by the TPG Rise Fund,[5] and the Anticipated Impact Measurement and Monitoring (AIMM) tool used by the IFC.[6] Each of these, along with others that have been developed more recently, has their pros and cons. The IMP, for example, is highly customizable and best for asking critical questions but also harder to quantify. The IMM, on the other hand, is more rigorous, quantifiable, and comparable with financial tools such as IRR or NPV but relies heavily on assumptions from a few academic studies and allows cherry-picking favorable supporting assessments without incorporating risks. The AIMM tool is also rigorous, considers the country/sector context of the investment, and highly structured in its approach but requires significant effort and resources to implement. Given these trade-offs, investors would be wise to learn from existing practices and adapt/customize their own approach accordingly.

Rather than go into detail on each of these methodologies, the more important, relevant, and practical question is: How can we improve, professionalize, and speed up the practice of measuring and managing impact throughout the investment process? The rest of this chapter explores six ways this can be done:

1. Redefine what we mean by "value."
2. Incorporate broader stakeholder perspectives to define what matters.
3. Increase standardization of impact measurement where appropriate.
4. Improve transparency and rigor of measurement where standardization is not appropriate.
5. Integrate financial and nonfinancial performance assessments.
6. Make impact reporting mandatory.

Let's discuss each in turn.

Redefine value

Our current ways of measuring value in society are flawed. In fact, many societal challenges—whether social, economic, environmental, or even political—boil down to the misalignment of incentive structures and human behavior created by our current ways of measuring value in society. For example, Gross Domestic Product, which is a central measure of value, has major limitations that lead to favoring production over human welfare and environmental sustainability. After reading Mariana Mazzucato's thought-provoking book *The Value of Everything*, I synthesized four primary flaws with our current measures of value and progress in society[7]:

Historical. Our System of National Accounts was created for the United Nations in the post–World War II era and overemphasizes production and military output and undervalues other activities. For example, war requires expanded arms and military forces, which increases production and thus GDP, but war is certainly not desirable. In today's era, where the statistical evidence shows life is getting better on a wide range of measures including health, peace, and poverty across the world, we should reconsider whether production really is the best or only measure of progress in society.

Misplaced production boundary. Even if we accept production as a measure of progress, our current system of accounting does not include home labor as output, misunderstands R&D's long-term impact on productivity, and ignores black markets, among other errors. I find particularly troubling the view that unpaid home labor does not contribute to the production and therefore has no value in society. The value parents provide by raising children is essential to a well-functioning society, yet this value goes unacknowledged by our economic system. Frustratingly, hiring laborers from outside the household to perform the same work would, in fact, be recognized as adding economic "value"! If we each made agreements with our neighbors to swap payments for housework, we would increase GDP overnight.

Happiness has no price. Increasingly, economists and other social scientists recognize that happiness is not merely a function of consumption, as classic utilitarian economics assumes. This is especially true once basic human needs and levels of material wealth are achieved. Rather, happiness also relates to behavior and quality of life, including elements such as social support, generosity, physical and mental health, and even number of days of sunshine or number of countries we can visit with our passports.

Unaccounted for externalities. Finally, and perhaps most importantly, our current systems of value measurement do not properly account for environmental externalities—such as air and water pollution—or their thresholds[8] and other side effects of production that current market systems cannot price. Environmentalists have long opposed capitalists because markets don't properly price for environmental damage, such as carbon emissions. This is a major weakness of our current system, and future generations will pay the price.

Further, when we look more closely at how we invest our money we also see the huge disconnect between financial markets and our values. The prevailing paradigm separates our investment decisions from our charitable decisions: Get rich first, and then give it away.[9] And despite the growth of the impact investing movement and its bold vision for the future, I have become increasingly convinced that we will never truly mainstream impact into investing and align long-term societal value interests with finance until society both reconsiders and redefines value *and* reimagines how it is taught in school and measured in business and government. In fact, impact investing will prove irrelevant if underlying systemic inequities and injustices continue to exist and go unaddressed because of deficiencies in how we define value and invest our money.

These problems are not intractable. We can do better by reconsidering what we value and how we make investment decisions. Below I suggest four mutually reinforcing concepts for how to refocus our investments on true long-term holistic value creation.

It starts by choosing to be 100 percent impact-aligned, or put differently, to make our values an explicit part of all our investment decisions. The accepted paradigm of investing based only on whether the opportunity will return capital at a higher rate than alternative options is no longer acceptable. We need to also consider whether what we define as important in our lives aligns with the investment. So, for example, if you have strong beliefs about nutrition or the need to use renewable energy, then you shouldn't be investing in companies and projects that make money but may also contribute to public health problems or continued use of nonrenewable energy sources. Some donor-advised funds, foundations, and family offices have started to be 100 percent aligned to impact (such as Ceniarth), but not enough of them. Being 100 percent aligned to impact mean that all of a foundation's assets are directed toward impact investments and none toward traditional capital preservation strategies.

Next, we need to redefine our customers and invest to address the pain points of not only consumers but also of society and our planet. Sustainable development goals have provided clear direction on helping investors understand which of the world's problems are the most pressing and in need of attention. Putting initiatives like SDG Impact[10] into practice gives investors a framework and provides indicators that help connect how investments contribute to solving the SDGs. More thinking of this type is needed.

Third, we need to invest in assets and enterprises that not only benefit stakeholders and avoid harm but that also proactively seek to contribute solutions to society's and the planet's problems. Beyond Meat provides a good example of this type of business. Beyond Meat offers a portfolio of revolutionary plant-based meats for consumers to experience the taste, texture, and other sensory attributes of meat while enjoying the nutritional benefits of eating plant-based foods. In doing so, it also has the potential to radically reduce carbon emissions by lowering the demand for traditional beef, one of the leading contributors to climate change.

Finally, we need to think long term when we make investments. Simply seeking to maximize return in the short term without thinking about an investment's long-term impact undermines the broader impact it can generate for society. Data and technology make it easier than ever to speculate and invest to "beat the competition." Yet that type of thinking not only creates winners and losers, but it also winds up hurting society and our planet. As extensive research has shown, the best investment strategy involves owning and investing in businesses that seek to create long term, sustained value for customers and society.[11] If investors thought long term, impact investing would become unnecessary, as all investments would create value for society.

We need to reopen the debate about how we measure value in business, finance, and society. One place to start is with how we invest and make investment choices. By choosing to invest for the long term, 100 percent based on our values, in organizations that proactively contribute solutions to people and the planet, we can be more certain our investments will truly have an impact.

Incorporate broader stakeholder perspectives

While not without its shortcomings, the Sustainable Development Goals (SDGs) and its predecessor, the Millennium Development Goals (MDGs), have been great at creating a unified set of targets to which all the numerous global governmental organizations can hold themselves accountable and progress can be tracked globally. However, the SDGs are primarily set top-down, much like goal setting within many impact investment organizations. Leaders opine on what they think is important, consulting with leading academics and experts, and then cascade those goals down through their organizations and to the ground in developing/emerging markets. Rarely, if ever, are the stakeholders who will be affected by these decisions consulted. This is a problem. If the people we are trying to benefit cannot identify and agree with the goals and targets, then those goals and targets are not useful.

In response, leaders of impact investment organizations should balance top-down goals and processes with a bottom-up and consultative process with the end beneficiaries and other relevant stakeholders, such as employees and communities affected or environmental agencies, to define impact goals. Further, annual strategic planning and goal-setting exercises should incorporate input from the realities on the ground as much as possible. For example, the Skopo Impact Fund defines goals based on end user experience, gets to know end users' personal stories, and uses those stores as feedback loops to refine goals continuously.

While incorporating diverse stakeholder perspectives is no doubt desirable, it can be difficult and costly in practice, as few private investors have the resources to do it and many stakeholders may not agree on what matters. These are still not sufficient excuses. Greater effort must be made to incorporate diverse voices. Otherwise, we will wind up continuing or even exacerbating the same problems and systemic inequalities for which we are seeking to solve.

Increase standardization of impact measurement

The end of the 1960s saw a series of controversial and inconsistent applications of financial accounting principles, with a flurry of mergers and acquisitions in the United States and the rise of companies' "opinion shopping," pitting accounting firms against each other. The result: a crisis in how financial accounting was used and interpreted. This crisis was the impetus behind the 1972 Wheat Study on Establishment of Accounting Principles and led to the establishment of the Financial Accounting Standards Board (FASB) on July 1, 1973.[12]

Today we are seeing a similar crisis concerning a different type of accounting. This crisis relates to the lack of consistently applied professional standards for impact measurement and management within the impact investing practice. Standards are lacking for aggregation of impact measures across different areas of investment or within areas of investment across organizations. Consequently, those seeking to compare impact investment funds must essentially rely on marketing claims without analytical substance. Most impact measures reported by impact investment organizations are, in fact, what I call "reach" (or output) indicators (e.g., products sold, jobs generated, or individuals trained), not what I call "depth" (or outcomes or impact) measures (e.g., how much incomes increased). Such approaches lead organizations to overprioritize investments achieving traditionally defined scale and to underprioritize investments with deep impact but without reaching scale.[13] They also often inadequately summarize the ecosystem-level impacts of different impacts, which can be difficult to quantify and observe within investment time periods.

For this reason—along with the need to measure progress against the SDGs as well as the increasing interest and entry of large traditional private equity firms into the space—improved, consistently applied professional standards for impact measurement and management are needed.

At the same time, vigorous debate has emerged within the impact investing field about the viability of attempting to standardize. There are many nuances specific to investing in different geographies, sectors, and asset classes. Even with recent advances in data collection and analysis, data availability and quality remain a constant challenge for anyone who has tried to do impact measurement and management. Even when data are available, different stakeholders will always have different preferences and goals, and there is no single "correct" way to prioritize among different preferences when making decisions. Further, different dimensions of impact may become more material at different moments, making it hard to plan when designing a single set of metrics.

I believe we will never reach a perfect solution *and* that, in general, improving impact measurement and management (IMM) standards will benefit enterprises and investors, much like standardization of financial accounting and reporting practices did in the 1970s and 80s. The industry and its funders should harmonize and adopt existing standards for the practice of impact measurement, aggregation, comparison, and communication. Existing resources should either be more widely adopted or improved. To start, there are many elements of IMM which could be standardized, including principles and classification systems. I am increasingly seeing the emerging market investor community coalesce around the SDGs, OPIM, and the IMP norms. As mentioned in the previous chapter, Cathy Clark at Duke University created a Coursera course to educate investors and entrepreneurs on how to implement IMM practices to achieve the SDGs. This and other similar efforts should be furthered.

Similarly, quantitative and qualitative indicators, metrics, and disclosures used by enterprises to measure, manage, and report their outcomes, and by investors to understand enterprises' performance, could be much more standardized. This could include a standard set of indicators for emerging versus developed markets as well as relevant metrics for different sectors or types of economic activity. Here too we have also seen a plethora of useful indicator databases emerge that are being used by a range of investors, including but not limited to the Global Reporting Initiative (GRI), Sustainability Accounting Standards Board (SASB), Harmonized Indicators for Private Sector Operations (HIPSO), CDP, and IRIS+. As I focus primarily on advising clients investing in emerging markets and have provided input to their development, I always advise investors to use GIIN's IRIS+. It is an incredible repository of relevant indicators with definitions and can be easily toggled for different use cases and investment themes.

Finally, we should further efforts to adjust/augment traditional financial accounting and reporting frameworks and tools for the variety of stakeholders and the broader impact on society that currently is unaccounted. A standardized framework to do that is being developed as of the writing of this book and is called the Impact-Weighted Accounts Framework (IWAF).[14] The overarching goal of IWAF is to guide organizations to create their own Impact-Weighted Accounts (IWAs). IWAs are a set of comprehensive quantitative and valued accounts containing impact information about the broader impact of an organization on its key stakeholders and the world including communities, the environment, and employees. IWAs consist of two key reporting accounts: (1) The Integrated Profit & Loss (IP&L) extends the "normal" P&L. It shows all impacts on stakeholders in one year; (2) the Integrated Balance Sheet (IBaS) extends the "normal" balance sheet. It shows impact assets and liabilities.[15] Once put into practice these accounts will fundamentally shift impact reporting for the better.

Improve transparency and rigor of measurement where standardization is not appropriate

The primary value of standardization is the comparability that it enables, whether that's within or across organizations. Yet, as previously discussed, it has its limits. There are places standardization of approaches may be counterproductive, namely impact rating methodologies and cutoffs on what constitutes a "good enough" level of impact for investment.

No single Impact rating methodology can be one-size-fits-all, and in some sense, the rating methodology is the secret sauce of how an organization makes investment decisions. Thus, it is difficult to mandate a single approach. Further, standardizing impact rating methodologies could erode the nuance in a variety of performance metrics and create a sense of false precision in measurement. Like financial valuation methodologies, investors should be at liberty to adapt the IMM norms and classification principles as they see fit. I consistently see my private equity clients who are starting impact investment funds do this. They start with the IMP norms and then adapt them to the specific theme of their funds, such as agriculture, climate change, energy, gender, or diversity and inclusion.

Similarly, it would be unwise to set standards on what is "good enough" in terms of impact performance. Different organizations have different goals, and what is important and of value to one organization will differ from others. Standardizing cutoffs of what is "good enough" risk homogenizing the importance of impact based on the opinions and perspectives of the experts with the money to invest. It could result in another form of impact colonialism and thus should be avoided.

In the areas where standards are not possible, we must increase the transparency and rigor by which impact is measured and performance is reported. Credible measurement approaches include independent randomized control trials (RCTs) or independent mixed-method surveys or studies. Unfortunately, many existing studies were generated or paid for by the industry and may have a self-justification bias. Additionally, many organizations lack the leverage or resources to evaluate their work, and others are uninterested in doing so. Where impact measurement does occur, it often yields to emotion, justifying activities and decisions rather than driving them. Further, organizations rarely demonstrate that their investments achieve the impacts promised. Microfinance exemplifies the danger of this failure. Some argue that the amount of capital available promotes learning by experiment, but the opportunity costs are too great. Using data and evidence to drive investment decisions is the exception, not the rule, and this needs to change.

This issue is not easy to solve. Incentives to focus on traditionally defined scale metrics should be reduced, and those favoring depth-of-impact focus, even if fewer people are reached, should be increased. Further, more soft funding should be made available to impact investment organizations to conduct independent evaluations of their work (including RCTs) and to promote needed revisions over time. One notable example of this type of effort is with the Global Innovation Fund, which explicitly looks for RCTs as part of its evaluation process of new investments while also recognizing they may not be appropriate for all situations. Finally, investors—both individuals and institutions—must increasingly hold impact investment organizations accountable to impact depth metrics or they will never change.

Integrate financial and nonfinancial performance

A further challenge many organizations face with IMM is the inability to understand and compare impact performance or integrate it with financial performance. Investment organizations have been using financial metrics like Return on Invested Capital (ROIC), Internal Rate of Return (IRR), Net Present Value (NPV), and others to evaluate investment opportunities before, during, and post-investment. Now with the emergence of impact classification systems and rating methodologies, they have options to evaluate and score investment opportunities from an impact lens. Yet these impact reporting tools may not always speak to each other and could in fact lead to inconsistent conclusions.

Further, for the most part, financial and impact analysis remain highly siloed in most organizations. Impact measurement specialists at investment organizations typically have their own teams with their own vernacular, frameworks, and data, all of which exist in varying degrees of isolation from their financial counterparts. I know this firsthand, as it was exactly how it worked at both the IFC and at Global Partnerships where I used to work.

Rather than comfortably pursue these silos, investors of all types should integrate their nonfinancial performance with their financial performance, at both the individual investment and portfolio level, fully embedded into the investment decision-making process. I find it most suitable for most private investors to start with the impact assessment and only proceed to the financial assessment if the impact reaches the desired impact threshold. This signals internally to staff and externally to potential investees and sources of funding that impact matters most and is taken seriously.

Impact Frontiers has developed a useful set of guidelines for investors on how to integrate financial and nonfinancial performance.[16] Essentially, it boils down to starting with an independent analysis of financial and impact performance but then looking at them together and holistically when making individual and portfolio-level decisions. This also means increasing the weight and importance of impact assessments pre-investment and not just making it a "check-the-box" exercise. Giving impact specialists investment veto rights or including the impact chair on investment committees with veto rights is a proactive way that many investors are doing this.

Finally, to the extent possible, investment organizations should seek ways to use hybrid metrics that link social and environmental performance with financial performance. Instead of treating impact performance and financial performance as two separate performance silos, it would be desirable to use metrics that make the material connection between the two explicit. Potential hybrid indicators like EBITDA/CO2 intensity for energy or EBITDA/yield per hectare for agriculture are two examples, but there are others. Hybrid metrics could help not only to improve ESG reporting but also allow easier analysis of performance that rewards companies that perform best in both social and financial dimensions.

Make high-quality impact reporting mandatory

As discussed in the opening of this chapter, the lack of standards for impact measurement, management, and reporting undermines the credibility of impact investing as a viable investment strategy. Most organizations continue to see impact performance reports as primarily useful to support fundraising and marketing efforts rather than for decision-making or to drive value creation.

Thus, LPs and other funders need to raise the bar by requiring their investees (GPs primarily) to provide both private and public impact performance reports. They need to also ensure a minimum standard for what should be included in an impact report and how frequently it needs to be reported. As the recent Tideline report highlighted, a good impact report needs to provide clarity and completeness[17]: *clarity* in that it presents relevant impact information in a manner that is digestible and facilitates analysis and decision-making, and *completeness* in that it provides comprehensive information from a broad spectrum of stakeholders at the individual and portfolio level needed to understand impact performance. A good report must also present and discuss any areas of underperformance as well as any impact risks and how they have been addressed.

An impact report should not just comply with standards and existing frameworks but also speak to the uniqueness of the impact thesis of each individual fund or investment. It should ideally present a clear theory of change and associated outcomes expected to be achieved by investment close before discussing performance. A standard practice is also to include case studies of specific investments with insights and perspectives from the relevant stakeholders involved in the investment.

Ideally, impact reporting is also fully integrated with financial reporting, in data collection, analysis, and reporting. During my work at Global Partnerships, following a lot of initial effort we were able to integrate our requests for both financial and impact data into a single automated portal that existing and potential investees could populate. Many DFIs are also following this approach, recognizing that if impact is to be taken seriously, it must be at the same level of importance as financial data.

The reports themselves then should also be developed and distributed in the same manner as financial reports, including organizing regular calls to discuss performance with investors and highlighting areas of both strong performance and unexpected underperformance. A good practice is to conduct quarterly calls that incorporate financially and impact updates and to have a separate annual impact and financial reports that go into much greater detail. At Dalberg, I have had the opportunity to work on numerous impact reports with large investors, and I am increasingly seeing this way of reporting as the standard and to be expected from impact-seeking LPs.

Summary of key messages from this chapter

- The main reason ESG and impact investing strategies are not being more widely adopted is due to lack of a common set of standards for how to account and measure nonfinancial performance.

- Six ways to raise the bar on impact measurement and management in impact investing are to (1) Redefine what we mean by value; (2) Incorporate broader stakeholder perspectives to define what matters; (3) Increase standardization of impact measurement where appropriate; (4) Improve transparency and rigor of impact measurement where it is not appropriate to standardize; (5) Integrate financial and nonfinancial performance assessments; and (6) Make high-quality impact reporting mandatory.

Notes

1. Zeisberger C (2022) Can Private Equity Make Money While Doing Good? INSEAD Responsibility Blog. https://knowledge.insead.edu/blog/insead-blog/can-private-equity-make-money-while-doing-good-18026. Accessed 27 July 2022.
2. BlueMark (2022) Raising the Bar: Aligning on the Key Elements of Impact Performance Reporting. https://bluemarktideline.com/raising-the-bar/. Accessed 28 July 2022.
3. Eccles R, Shandal V, Young D et al. (2022) Private Equity Should Take the Lead in Sustainability. Finance & Investment, Harvard Business Review. https://hbr.org/2022/07/private-equity-should-take-the-lead-in-sustainability. Accessed 30 July 2022.
4. Learn more by visiting: https://impactfrontiers.org/norms/.
5. Learn more by visiting: https://hbr.org/2019/01/calculating-the-value-of-impact-investing.
6. Learn more by visiting: https://www.ifc.org/wps/wcm/connect/topics_ext_content/ifc_external_corporate_site/development+impact/aimm.
7. Mazzucato M (2018) *The Value of Everything: Making and Taking in the Global Economy*. PublicAffairs, New York.
8. Impact Management Platform (2022) Thresholds and Allocations. https://impactmanagementplatform.org/thresholds-and-allocations/. Accessed 29 Aug 2022.
9. Miller, Clara (2016) Building a Foundation for the 21st Century. *Nonprofit Quarterly*, 8 June 2016.
10. Find out more by visiting: https://sdgimpact.undp.org/.
11. McKinsey (2020) *Valuation: Measuring and Managing the Value of Companies*. Wiley.
12. Zeff S (2015) The Wheat Study on Establishment of Accounting Principles (1971–72): A Historical Study. *Journal of Accounting and Public Policy* 34(2): 146–174. https://doi.org/10.1016/j.jaccpubpol.2014.12.004.
13. Scale should be considered both in terms of breadth, e.g., how many people benefit and depth, e.g., how much each individual benefits from each intervention.

14. Find out more by visiting https://impacteconomyfoundation.org/impactweightedaccountsframework/.
15. Ibid.
16. Impact Frontiers (2020) Impact-Financial Integration: A Handbook for Investors. https://impactfrontiers.org/wp-content/uploads/2022/05/Impact-Frontiers-Impact-Financial-Integration-A-Handbook-for-Investors-Updated-July-14-2020.pdf. Accessed 1 Aug 2022.
17. BlueMark (2022) Raising the Bar: Aligning on the Key Elements of Impact Performance Reporting. https://bluemarktideline.com/raising-the-bar/. Accessed 27 July 2022.

10

"Mainstream Impact Management"—An Interview with Olivia Prentice

This chapter summarizes my interview with Olivia Prentice, Head of Impact at Bridges Fund Management and former head of content and COO at the Impact Management Project. Through that work she led the development of the now widely adopted and recognized standards of impact measurement and management in the field.

Question 1: What is the Impact Management Project and how did it first start?

Olivia: The Impact Management Project (IMP) began in 2016 as a time-bound forum for building global consensus on how to measure, assess, and report impacts on people and the natural environment.

The IMP was started in response to the challenge—experienced by many investors, corporates, standard-setters, and funders alike—that there was no consensus on how to measure, improve, and disclose sustainability impacts, and that different language and approaches were being used across the value chain, which also led to a proliferation of standards and frameworks.

IMP was a grant-funded initiative with a lean, small team which benefited from a group of generous advisors and large community of experts. List of funders/advisors can be found on our website.

From 2016 to 2018, the IMP brought practitioners of all kinds together to share their experiences managing impact to identify approaches common to all, including the dimensions of performance data needed to understand

K. Hornberger, *Scaling Impact*,
https://doi.org/10.1007/978-3-031-22614-4_10

a single impact on people or planet. (Per the above note, this intentionally spanned many geographies, disciplines, and perspectives).

This was not a new framework, but a set of norms identified by looking at existing data, reports, and language and by listening to how stakeholders globally talk about their lived experiences. These dimensions can be used as a checklist to ensure the information needed to understand any social or environmental impact is present. This work led to the 5 dimensions of impact.[1] (*See below for more information on how these 5 dimensions enabled a logic for also identifying archetypes of company impact, and therefore classes of investment.*[2])

To ensure the norms were providing a helpful grounding for practitioners, we did a lot of market outreach and engagement work to facilitate market comprehension of the norms, and get feedback from different corners of the world about how they were used in practice. For example, during 2018–2020 we facilitated debate on new topics via the HBR Idea Lab and published papers illustrating emerging consensus on more nascent areas of practice, such as impact valuation and investor contribution measurement.

We noticed that investors in particular were finding the 5 dimensions/impact classes helpful for articulating their impact goals and identifying what information they needed to collect to evaluate performance. However, investors and companies were still struggling with how to use the range of standards, tools, and frameworks that were emerging. At the time, there was a lot of frustration about the "alphabet soup" of standards, and an assumption that all standards/frameworks were trying to do the same thing and were therefore duplicative or interchangeable. It felt important to have the standard-setters come together to explain that different standards have different purposes and move toward greater interoperability between them.

With this vision, from 2018 to 2021, the IMP facilitated standard-setting organizations—called the IMP Structured Network—to clarify the landscape of standards and guidance used by practitioners for their impact management practice. The goal was to—for the first time—have an explanation from the standard-setters themselves about how their standards differ but also about how the various standards can be used in combination to form a complete practice of impact management (this is what the Impact Management Platform now contains).

The standard-setters collectively decided to use some shared "Actions of impact management" as the basis for explaining the interoperability of their standards. They hoped to show that different standards are needed to execute each action of impact management. Sometimes there were many standards catering to the same action, sometimes there were none, but the main challenge was a lack of connectivity between actions. For example, the metrics used to measure impact weren't always those used by the benchmarking tools.

Through this work we were able to support the development of a tool that can be used by any enterprise or investor, to find the resources needed to manage sustainability impacts. The Impact Management Platform represents an evolution of the Structured Network facilitated by the Impact Management Project from 2018 to 2021. With a Steering Committee that brings together a set of multilateral,[3] the Platform is a collaboration that enables ongoing coordination among leading international providers of sustainability resources.

Its work program is designed to mainstream the practice of impact management, including identifying opportunities to consolidate existing sustainability resources, collectively addressing gaps, and coordinating with policymakers and regulators.

To provide the clarity that practitioners have been calling for, the Platform created a regularly updated web tool that explains the core actions of impact management and links to the resources that support organizations and investors to implement them. The Impact Management Platform is an essential "go to" for practitioners that want to practice impact management.

Question 2: What are other impact management mainstreaming activities IMP was involved in?

Olivia: Alongside standard-setting the dimensions of impact, IMP supported three other initiatives that are collectively well-positioned to mainstream impact management, so that all organizations have the resources they need to improve their sustainability impacts:

First, we sought to integrate impact into investment practices—Impact Frontiers is a peer-learning and market-building collaboration, developed with and for asset managers, asset owners, and industry associations. The initiative creates practical tools and peer-learning communities that support investors in building their capabilities for managing impact, and integrating impact with financial data, analysis, frameworks, and processes. Impact Frontiers also facilitate further consensus-building in areas of practice where standards and guidance do not yet exist, using practitioner experience to jump-start the conversations.

In addition, we sought to set a global standard to report impacts that inform investment decision-making—In November 2021 the International Sustainability Standards Board (ISSB) was launched to develop—in the public interest—a comprehensive global baseline of high-quality sustainability disclosure standards to meet investors' information needs. The ISSB sits alongside and connects to the International Accounting Standards Board (IASB), under the governance of the IFRS Foundation and its monitoring board of public authorities from jurisdictions around the world.

To achieve that goal in 2020, the IMP facilitated the leading sustainability and integrated reporting initiatives (CDP, CDSB, GRI, and the VRF) to co-create a shared vision for corporate reporting, as well as prototype climate and general disclosure requirements to provide the IFRS Foundation with a running start for the development of international sustainability standards by the ISSB. Then in March 2021, the IMP's Chief Executive, Clara Barby, took partial leave from the IMP to be the project lead for the IFRS Foundation's development of the ISSB, under the oversight and strategic direction of the IFRS Foundation's Steering Committee of Trustees.

Finally, we sought to assess and disclose the impact classes of investments—The Impact Classification System (ICS) is a self-assessment and reporting tool for investment practitioners wanting to disclose how and to what degree their financial products meet sustainability goals. The ICS utilizes the impact classes developed through the IMP's consensus-building efforts. Launched in June 2020, the ICS aggregates information about sustainability-related disclosures of investments and brings them into a consistent digital reporting format. The ICS was developed through the IMP+ACT Alliance, a two-year technology project seeded by the IMP and Bridges Insights. Following the successful rollout to over 300 practitioners, the GIIN has now brought the ICS into its suite of impact management offerings, and is enabling the ICS to remain freely available. Interested parties can register for access through the GIIN website.

Five years after its launch, having achieved these deliverables, the Impact Management Project has concluded as planned. More details on what comes next below.

Question 3: How would you define impact in the context of investing for impact?

Olivia: Through the Structured network of 18 standard-setters[4] we agreed on the following definition of impact: "A change in an aspect of people's well-being or the condition of the natural environment caused by an organization." An enterprise has many impacts, positive and negative. For investors, the impact of the enterprise needs to be considered alongside investor contribution.

The IMP explicitly did not try to define "impact investing" given definitions already existed, and in addition it might be that different investors might point to different impact classes to convey their expectations of "impact investing," so agreeing on the impact classes themselves was what we focused on.

Question 4: How did you identify and land on the IMP five dimensions of impact?

Olivia: They were developed through a lot of consultations with a wide range of stakeholders. We brought together and held consultations with over 3000 organizations including a range of investors (ethical, ESG, impact, fiduciary, concessionary), companies (SMEs, multinationals, nonprofits), and policy-makers, alongside academics and standard-setters. The idea was to find a set of language and "norms" for impact management that everyone could agree on. We complemented that by observing dimensions of impact measured and data collected by companies and investors already, as well as by looking at existing standards such as SVI and GRI. We then spent a lot of time observing and listening to the lived experience of stakeholders globally.

Question 5: What is the IMP classification system and how can investors use it?

Olivia: A description of the classification system can be found on the Impact Frontiers website[5] but to summarize I would say the following: A growing number of investors are motivated to manage the effects of their investment portfolios on people and the planet. These investors vary widely in their intentions and constraints. For example, a passive retail investor looking to mitigate risk by avoiding harmful activities is likely to construct a different investment portfolio for a foundation using its entire capital base to advance its charitable mission. Different again is an institutional investor's portfolio that anchors new investment products addressing social issues in its clients' communities.

To align their portfolios with their intentions, all investors need to be able to understand the actual impacts of the assets or investment products/funds available to them. In addition, the intermediary investment managers and the assets seeking investment want to identify aligned investors and avoid being compared inappropriately to assets with different impact goals, or avoid being judged on just financial performance alone.

Impact classes group investments with similar impact characteristics based on their impact performance data (or, in the case of new investments, their impact goals). Impact classes bring together the impact performance (or goals) of the assets being invested in and the strategies that the investor uses to contribute to that impact. They can be used to define boundaries within which comparisons of impact performance are likely to be possible and sensible.

Impact classes are not intended as a replacement for progress toward a global performance measurement standard that could enable the impact of individual investments to be compared. Instead, impact classes offer an immediate and complementary solution for differentiating the type of impact that investments have, even when very different measurement approaches are used (e.g., reports vs ratings vs certifications).

Investments can be classified on the Matrix in two ways:

- An investment's (or portfolio of investments) impact goals. For most investors this will involve selecting one or two impact classes which align with the impact they expect or intend to contribute to through their portfolio. For other investors, the range of the strategy could be broader and cover multiple impact classes. Where possible, investors are encouraged to indicate what proportion of assets under management they expect to allocate into each impact class, noting allocation thresholds where they exist.
- An investment's (or portfolio of investments) actual performance. When plotting the actual impact of investments on the matrix, investors are encouraged to indicate what proportion of AUM is allocated into which impact class(es). To be transparent, and ease impact management decisions for their stakeholders, investors should display this information alongside their original goals. Where relevant and possible, investors can then explain to their stakeholders where and why performance might differ from the original goals.

Question 6: How would you encourage investors to incorporate the IMP norms in practice?

Olivia: I would only encourage the use of the IMP norms in addition to existing standards. For example, the 5 dimensions and other normative practices identified through the IMP are now baked into many standards including aspects of ISSB disclosure requirements, SVI, UNDP, and IRIS+.

The impact management norms are best thought of as building blocks. If you are starting from scratch, you may want to build your impact management framework based on them. If you already have an impact management framework, you may want to use them as a checklist to ensure that you are not missing any essential elements.

Though not all investors have goals with regard to all five dimensions, it is likely that someone up or down the capital chain does, so it is recommended wherever possible to measure and disclose goals and performance against all five.

Question 7: Why is it important to integrate financial and impact performance?

Olivia: This is an incredibly important question and the focus of our new host organization Impact Frontiers and its Executive Director, Mike McCreless. To quote Mike he would say: "Essentially impact and financial performance should be reinforcing, and thus it is important to integrate the two to make sure they are each valued. This should happen both at the transaction and the portfolio level through a set of harmonized metrics and practices.

This starts by creating a measurement tool to aggregate the key impact performance elements of an investor's impact thesis, ideally tied to the five dimensions. Metrics such as the number of people reached, how underserved those people are, and how much everyone is affected should be somehow aggregated via a customized methodology. This impact aggregate metric such as a rating should then sit alongside a measure of financial return such as internal rate of return (IRR), net present value (NPV), or return on invested capital (ROIC).

The two should then be viewed together both when making decisions about individual investments as well as on aggregate across the portfolio, almost as if on a 2 × 2 matrix that shows impact performance on one axis and financial performance on the other. This combined view allows for interesting discussions and dynamic decision making. Many of the organizations we worked with at IMP, including Bridges Fund Management, do this."

Question 8: What is next for the IMP now that it has transitioned to the Impact Frontiers platform?

Olivia: The IMP has concluded as planned. We believe it's the right time for the market and its standards to take forward the agenda of further coordination and consolidation.

IMP's conclusion does not mean the norms are not relevant or going away. On the contrary, they are embedded into multiple standards and on the Impact Management Platform website.[6]

There is of course more work to do; IMP is handing off a set of resources developed by the IMP Practitioner Community to Impact Frontiers, as well as the baton to continue to extend and deepen this consensus in areas such as investor contribution and impact portfolio construction.

Question 9: How do you hope the impact investing landscape will evolve over the coming years?

Olivia: I'd love to see impact management becoming mainstream, where every company and investor considers the negative and positive impacts of their activities and manages them within the constraints of their financial goals. Of course, it feels increasingly critical that more capital flows to the contribute solutions or "C" classification (which some would define as "impact investing") to find new solutions to the growing social and environmental challenges of our time.

I am hopeful that the recognition that these types of "C" investments can present attractive commercial returns continues to grow and strengthen as activity in the space continues to scale. I think we're seeing this potential with the explosion of new "specialist" funds that don't call themselves impact funds but are just that. And, alongside this, that we see flexible capital provision across every geography to fund solutions that fiduciary capital can't in its present form. Perhaps over time, we can get more creative with our investment products and allocations to broaden the scope of what fiduciary investors can invest in across a portfolio approach. I am hopeful that current trends will continue and in doing so we will be able to bring impact to scale.

Notes

1. Learn more by visiting: https://impactfrontiers.org/norms/five-dimensions-of-impact/.
2. Learn more by visiting: https://impactfrontiers.org/norms/investment-classification/.
3. To learn more about IMP governance visit: https://impactmanagementplatform.org/about/.
4. See list on the www.impactmanagementplatform.com website.
5. Learn more by visiting: https://impactfrontiers.org/norms/investment-classification/.
6. Learn more by visiting: https://impactmanagementplatform.org/.

Part IV

Offer Catalytic Finance, Not Just Blended Finance

11

Unleashing Private Capital for Global Development

As I mentioned in the introduction, the global community has committed to meeting an ambitious set of 17 Sustainable Development Goals (SDGs) by 2030—a wide range of goals including SDG 1, "End poverty in all its forms everywhere," SDG 3 "Ensure healthy lives and promote well-being for all at all ages," and SDG 13, "Take urgent action to combat climate change and its impacts." SDG targets include eradicating extreme poverty, ending preventable deaths of newborns and children under age five, ensuring universal access to quality education, and providing universal access to affordable modern energy services, among many others.

However, recent estimates of progress toward attaining these targets present a sobering picture. For example, since the beginning of the COVID-19 pandemic researchers estimate that there has been an increase of between 76 and 95 million people living in extreme poverty. Further, the pandemic is threatening decades of progress in global health decreasing life expectancy and basic immunization coverage and increasing the prevalence of anxiety and depression and deaths from tuberculosis and malaria. Worse almost 150 million children have missed in-person instruction from 2020 to 2021 and 24 million learners are expected never to return to school.[1]

If we are to meet global SDG targets, we will need significantly more innovation—from breakthroughs in technology to reimagined service delivery models to new forms of financing. The Commission on Investing in Health (CIH) found that even if today's health interventions—including medicines,

K. Hornberger, *Scaling Impact*,
https://doi.org/10.1007/978-3-031-22614-4_11

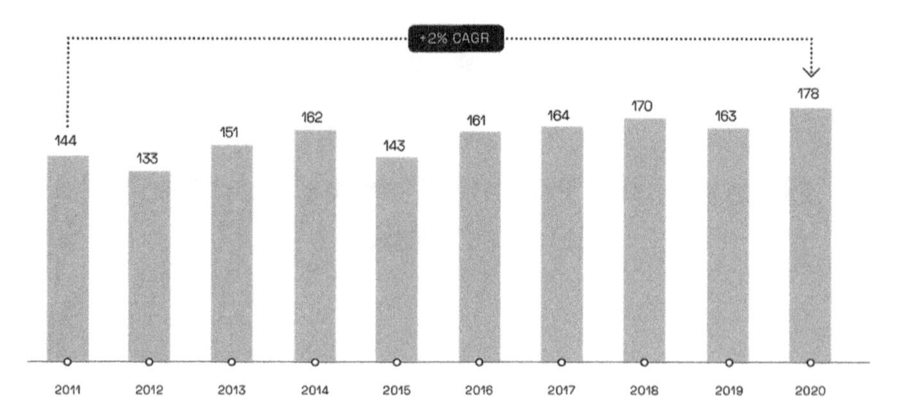

Fig. 11.1 Official development assistance (ODA) flows to developing countries ($ billion), 2011–2020 (*Source* OECD data focusing on I.A + I.B ODA flows to developing countries, author analysis)

vaccinations, bed nets, and diagnostics—were scaled up to 90 to 95 percent coverage worldwide, we would still fall short of many SDG3 targets.[2] Existing approaches are simply insufficient to expand access to high-quality, affordable education, energy, and healthcare for all—particularly the most vulnerable at the base of the pyramid (BOP).

We will also need new approaches to development finance, including more effective ways to mobilize private sector capital to support the innovation needed in education and global health and new and scaled methods to address climate change.

The gap in financing to meet the 17 SDGs stands at an estimated $3.7 trillion annually.[3] While only a quarter is for social infrastructure where private capital can play a role—the resource gap is still large.[4] Yet from 2011 to 2020, official development assistance (ODA) to developing countries has largely remained stagnant (see Fig. 11.1). Coupled with insufficient government spending on public services and infrastructure, this has resulted in enormous annual investment gaps in low- and middle-income countries (LMICs). Successfully attracting private capital to support development goals and achieving sustainable and inclusive growth, where commercial returns are also possible, is crucial to filling these gaps.

The opportunities and challenges of financing SDGs

We have good reason to believe both objectives are possible: that we can successfully increase the number of promising solutions that reach scale to address global challenges *and* that we can attract more private capital to play a role in supporting innovators on that journey.

On the innovation side, thousands of innovators are emerging in diverse settings—from academic and research institutions to corporate research and development divisions to local communities across the developing world. When these innovators sustainably scale, they can have a transformative impact in delivering quality products and services to BOP populations. To cite just a handful of examples: The Aravind Eye Care System has performed more than five million low-cost cataract surgeries to date in India[5]; an estimated 450 million cases of malaria have been prevented because of PermaNet long-lasting insecticide-treated bed nets (LLINs)[6]; and another innovator, GreenLight Planet, has provided solar lamps and home systems to over 82 million off-grid and under-electrified consumers, primarily in Sub-Saharan Africa.[7]

On the private capital side, significant capital remains on the sidelines. Mobilizing just 2.5 percent per year of the roughly $100 trillion AUM held by private asset managers would fill the entire financing gap for all 17 SDGs.[8] Perhaps more importantly, investors are increasingly interested in deploying private capital for impact. The total US-domiciled AUM using environmental, social, and governance (ESG) criteria has grown from $2 trillion in 1999 to $12 trillion in 2018—more than one-quarter of all professionally managed assets at that point in time.[9]

Impact investing has been buoyed by Millennials, 61 percent of whom, have made some form of impact investment, compared with just 23 percent of affluent donors in the Baby Boomer and older generations.[10] This tailwind is set to continue as most next-generation financial advisors (under 40 years old) are personally interested in impact investing, compared to only 52 percent of advisors over 40 years old.[11]

However, private investment in necessary innovations—even those with commercial potential—remains nascent today. Investors cite several unique barriers to investing in innovators focused on emerging markets that inhibit increased deal activity. First, the commercial markets remain relatively underdeveloped, given the often longer lead times for approvals and uncertainty around end customers; this reduces available deal flow. Investors also remarked that the level of required domain and technical expertise, as well as local regulatory knowledge, is comparatively higher in some sectors needing investment, such as healthcare, housing, water, and sanitation—just as it is in developed markets. Finally, investors note that the cost and complications of implementing a known and viable technology can be more difficult in emerging markets, all making investment more challenging.

The challenges to mobilizing more private sector capital for global innovation also occur on the innovator side. Promising innovators often stall or fail at each stage of the journey from initial idea to scale for various reasons (see Fig. 11.2). As a result, many never reach the minimum threshold of commercial viability to accept private sector capital. In other words, investors do not perceive sufficient innovator deal flow with risk-adjusted return profiles to be attractive relative to other investments they could make. Additionally, there is a mismatch between the types of capital available (return expectations, duration) and the capital needs of promising innovators at each stage of the innovator journey (see Fig. 11.2). As a result, not enough promising innovators reach a sustainable scale, limiting their potential impact in extending high-quality, affordable products, and services to the base of pyramid populations that need them.

Blended finance offers one promising avenue to fill the gap in global investment. Blended finance refers to the strategic use of development finance and philanthropic funds to mobilize private capital flows to emerge and frontier markets.

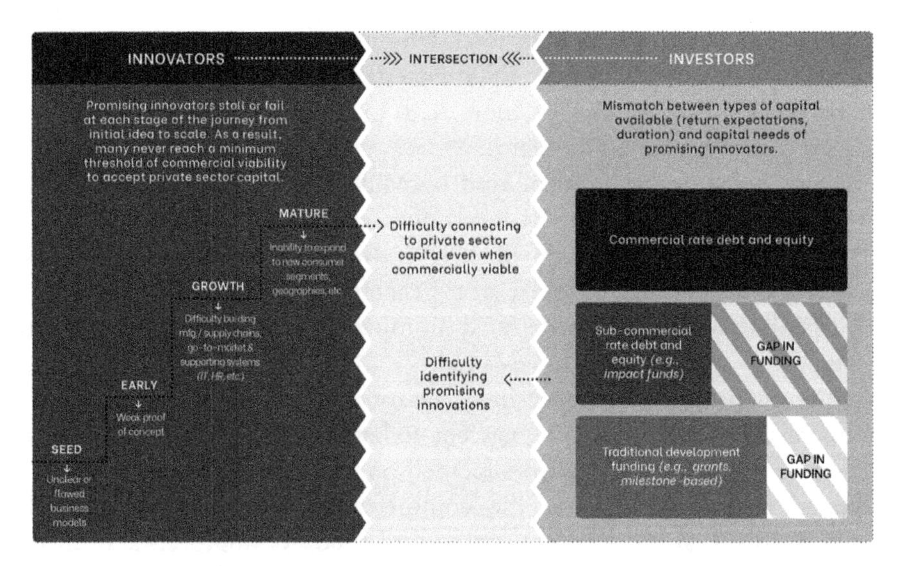

Fig. 11.2 Challenges for connecting investors' and innovators' investment needs

Development and/or philanthropic funding can be used to de-risk investment and improve the overall risk-adjusted return of emerging market investments, bringing it in line with investor expectations. Blended finance makes use of existing financial instruments and can focus de-risking on either side of the innovator and investor divide. For example, targeted technical assistance and catalytic grant capital can be provided to promising innovators who are just under the threshold of minimum viable business model sustainability; by allowing such innovators to refine strategies and begin to gain revenue or income traction, such support reduces perceived business model risk relative to the potential return.

On the other side of the spectrum, development funders can provide de-risking instruments directly to funds at the fund or deal level in the form of guarantees, junior equity, or subordinated debt; by providing a tranche of capital with asymmetric exposure to risk—typically capped returns and/or first-loss capital—development funders can help mechanically bring risk-adjusted returns in line with a wider set of private investors' expectations. I dive further into the use cases of blended finance in Chapter 12.

Potential to mobilize private capital

For development and philanthropic funders, blended finance offers leverage: a clear multiplier effect for every development dollar invested. Based on a sample of 72 different blended finance funds or fund-like structures, development funders were able to "crowd-in" approximately $4.10 of private sector capital for every $1 of development funding.[12] Perhaps not surprisingly given investors' increasing interest in investing for impact and the possibility of receiving returns near or in line with commercial rates, blended finance is rapidly gaining traction—with over $160 billion in financing mobilized to date.[13]

I believe a significant opportunity exists to enable more private capital flow to needed agriculture, climate, education, and health innovations through targeted development and philanthropic funder efforts to stand up new blended finance facilities and services. Indeed, if current annual growth rates continue, private capital mobilized through blended finance will total US$252 billion by 2030.[14] Even if the total allocated to the health sector remains at 5 percent, this represents an additional $13 billion for health funding alone.[15]

Making blended finance more "catalytic"

The term "catalytic capital" has generated considerable buzz in the impact investing and philanthropy communities in the last few years, largely due to a series of grants from the Catalytic Capital Consortium (C3), the organization that coined the term and that has subsequently worked to demonstrate and support the use of the concept.[16]

Yet the approach behind the term remains poorly understood—specifically, how it relates to other types of blended finance with philanthropic or donor support. As Chris Jurgens mentioned earlier C3 defines catalytic capital as "capital that accepts disproportionate risk or concessionary returns to generate positive impact and to enable third-party investment that otherwise would not be possible." And although this makes a catalytic capital sound akin to blended finance, the approaches differ in essence.

While all blended finance approaches use catalytic capital, not all uses of catalytic capital can be considered blended finance. Convergence, the leading intelligence platform on blended finance, considers a transaction to be blended only when catalytic capital attracts private financing into the structure's capital stack—thus making catalytic capital a key aspect of blended finance. An example of catalytic capital that is not considered blended finance is a 100 percent private venture capital investment into a new economic activity such as alternative proteins that results in follow-on investment but without any blending.

It's worth noting that Convergence does not account for or consider other types of "catalytic" activities, such as demonstration effects or facilitating innovation. Using the C3 definition and looking at the Convergence database of historical blended finance deals, we see that use of catalytic capital has achieved an annual growth rate of 26 percent over the last twenty years (see Fig. 11.3). In 2020, of the 54 blended finance deals, 72 percent included concessional debt, guarantees, or junior equity to facilitate and bring the deal to life.

Another mistaken belief about catalytic capital is that it comes only from private foundations or donors looking to de-risk others in the transaction. Few realize that a large share of catalytic capital also comes from development finance institutions (DFIs).[17] In 2020 alone, $303 million (or 28 percent) of concessional financing was provided by DFIs to de-risk or sweeten 24 deals for other investors (see Fig. 11.4).

So how do DFIs provide catalytic capital, and what might this mean for the future growth of its role in DFIs' plans moving forward? Some quick insider research reveals three "archetypes" for DFI's use of catalytic capital

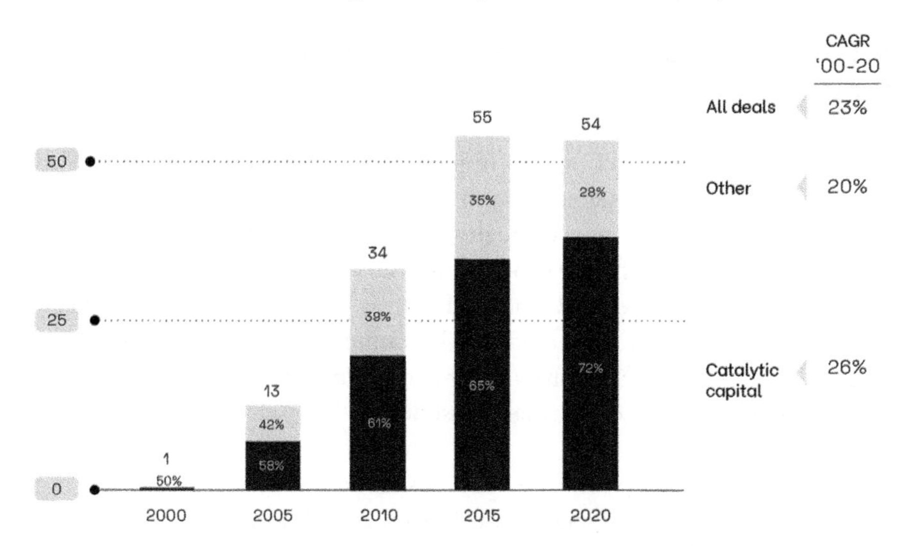

Fig. 11.3 Growth of blended finance deal volume, by use of catalytic capital and other blending approaches, 2000–2020 (*Source* Convergence historical deals database, accessed on 09/20/2021, Author analysis)

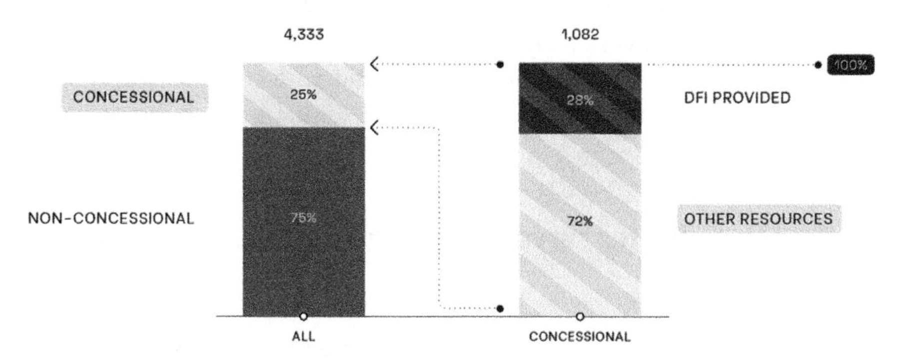

Fig. 11.4 Blended finance deal value in 2020, $ millions (*Source* Convergence historical deals database, accessed on 09/20/2021, Author analysis)

that I believe are worth sharing so others may learn from them and replicate their results: Demonstrate, De-risk, and Enhance Impact. Each is described below with examples.

#1: Demonstrate the potential of a new investment thesis

In 2016, Pomona Impact, a new impact investing fund was created with a focus on investing in Guatemala and Central America. As a first-time fund manager, the founders had difficulty raising capital. Pomona invested $1 million of friends' and family's money and was able to obtain a catalytic loan of $1 million from the IDB Lab (a DFI) at favorable terms with concessionary rates. The IDB Lab's loan was instrumental in getting the pilot fund to critical mass (of $2 million) and provided Pomona with the opportunity to demonstrate that it could make sustainable "impact investments" in the region.[18]

Two years later, the pilot fund was considered a success, investing in organizations like EcoFiltro, Uncommon Cacao, and Yellow Pallet. Based on the demonstration effect that impact investing in Central America was feasible, the Pomona Impact team has now successfully raised and launched its first $20 million fund, currently deploying capital into high-impact social enterprises across Central America with more than a dozen investors, including DFIs and private sources of capital. As illustrated in this example, DFIs can provide an early-stage private enterprise or fund with concessionary finance, that allows the asset to grow and later attract capital from investors seeking market-rate returns.

#2: De-risk to better serve the needs of difficult to serve social ventures

Traditionally, due to perceived and real risks, agriculture has been considered a difficult segment to finance. My own analysis while working at Dalberg shows that providing loans under $1 million to agriculture SMEs (ag-SMEs) is difficult to do profitably. We also know, however, that safe, reliable, environmentally sustainable food systems are critical to our shared prosperity. To respond to this urgent situation, Incofin, a fund manager based in Belgium, has introduced the Fairtrade Access Fund (FAF) to serve the financing needs of smallholder farmers and ag-SMEs by investing in Fairtrade-certified organizations and cooperatives in Latin America and Africa.[19]

To make the fund financially viable and to attract private investors like pension funds, commercial banks, and corporates, the FAF used an innovative blended structure with two share classes and the possibility of leveraging up to one time. These two share classes attract investors with differing profiles: Class A shares have a seven-year lock-up period, thus attracting investors with a long-term development focus (principally DFIs); Class B redeemable shares attract private investors.

The FAF has been a great success, having now made $128 million in disbursements and impacting more than 327,000 smallholder farmers' livelihoods through 10 percent increases in productivity and $250 income increases per family. Without initial junior equity financing from the FMO (the Netherlands' DFI), these investments—and thus these farmers' livelihoods—might not have had access to those opportunities.

#3: Enhance the impact potential of investments with technical assistance

The Dutch Good Growth Fund (DGGF) is publicly funded by the Dutch Ministry of Foreign Affairs (MFA) and privately managed by Triple Jump and Price Waterhouse Coopers (PWC). Since 2014, DGGF has been providing a combination of catalytic capital, capacity building, and ecosystem development to bridge the financing gap for SMEs in emerging markets. DGGF provides catalytic capital via its Fund of Funds (FoF) to the mezzanine, private equity, and venture capital vehicles, and concessional debt to upscaling microfinance institutions and digital lenders (to name a few examples) in a manner similar to that of many DFIs. Between 2015 and 2020, DGGF committed €324.8 million via its FoF to 60+ local finance providers and €37.6 million via its Seed Capital and Business Development (SC&BD) facility to 20 local finance providers and 10+ entrepreneur support organizations. The SC&BD facility aims to strengthen the mission impact of the DGGF investment funds by providing technical assistance (feasibility and market studies) and enterprise support/business development services to the intermediaries in their portfolios. They also have built a knowledge and research team that helps synthesize and share learnings across the portfolio on the use of catalytic capital. Combined, these additional services enhance and disseminate information on the impact of DGGF and its investments.

As these examples show, it is not only possible but critical that DFIs continue to channel donor resources and provide catalytic capital. DFIs that invest concessionally and prioritize market building and innovation can achieve the greatest impact. If a DFI hesitates to take this route, one natural means of scaling up is to pair the activities of bilateral donor aid agencies with bilateral DFI support. One notable example is the USAID INVEST program, which has been working with the US's DFI, DFC, via Prosper Africa—but there are others. Let's hope the growth trend in DFIs' use of catalytic blended capital continues.

Summary of key messages from this chapter

- Attracting more private capital to complement government and philanthropic sources of funding is critical to achieving the United Nations Sustainable Development Goals (SDGs).
- Critical challenges make it difficult for private investors to deploy capital in emerging markets to support innovations that will accelerate the achievement of the SDGs. These challenges include a perceived lack of sufficient domain expertise about investing in emerging markets or in sectors of needs, insufficient pipeline of scaled and commercially viable investment opportunities, and a mismatch in the type of capital available and the capital needs of enterprises at different stages of growth.
- Blended finance offers one promising avenue to fill the gap in global investment. Blended finance refers to the strategic use of development finance and philanthropic funds to mobilize private capital flows to emerge and frontier markets.
- Catalytic capital is a subset of impact investing that addresses capital gaps left by mainstream capital, in pursuit of impact for people and planet that otherwise could not be achieved. It accepts disproportionate risk and/or concessionary return to generate a positive impact and enable third-party investment that otherwise would not be possible.
- To date too little blended finance has been used to truly catalyze private investment that would not have happened otherwise—particularly the capital provided by DFIs. To make DFI investments more catalytic, they can use their blended finance windows to demonstrate the potential of new investment theses, de-risk to better serve the needs of difficult to serve social ventures, and enhance the impact of potential investments with technical assistance.[20]

Notes

1. United Nations (2022) The Sustainable Development Goals Report 2022. https://unstats.un.org/sdgs/report/2022/The-Sustainable-Development-Goals-Report-2022.pdf. Accessed 30 Aug 2022.
2. Boyle C, Levin C, Hatefi A et al (2015) Achieving a "Grand Convergence" in Global Health: Modeling the Technical Inputs, Costs, and Impacts from 2016 to 2030. The Commission on Investing in Health. *PloS One* 10(10).
3. OECD (2022) Closing the SDG Financing Gap in COVID-19 Era. https://www.oecd.org/dev/OECD-UNDP-Scoping-Note-Closing-SDG-Financing-Gap-COVID-19-era.pdf. Accessed 09 Sept 2022.

4. Pegon, Matthieu (2022) Mobilizing Beyond Leverage: Exploring the Catalytic Impact of Blended Finance. IDB Invest.
5. Ravilla T, Ramasamy D (2014) Efficient High-Volume Cataract Services: The Aravind Model. *Journal of Community Eye Health* 27(85): 7–8.
6. Hillebrandt H (2015) Bednets Have Prevented 450 Million Cases of Malari. Giving What We Can. https://www.givingwhatwecan.org/blog/bednets-have-prevented-450-million-cases-of-malaria. Accessed 3 Aug 2022.
7. Njanja A (2022) Sun King Raises $260M to Widen Clean Energy Access in Africa, Asia. Tech Crunch. https://techcrunch.com/2022/04/27/sun-king-raises-260m-to-widen-clean-energy-access-in-africa-asia/. Accessed 3 Aug 2022.
8. US SIF Foundation (2018) 2018 Report on US Sustainable, Responsible and Impact Investing Trends.
9. Ibid.
10. Fidelity Charitable (2022) Using Dollars for Change: Seven Key Insights into Impact Investing for 2022 and Beyond. https://www.fidelitycharitable.org/content/dam/fc-public/docs/insights/impact-investing-using-dollars-for-change.pdf. Accessed 30 Aug 2022.
11. Fidelity Charitable (2021) Impact Investing on the Rise: How Financial Advisors are Adapting. https://www.fidelitycharitable.org/content/dam/fc-public/docs/insights/impact-investing-how-financial-advisors-are-adapting.pdf. Accessed 30 Aug 2022.
12. Convergence (2018) Leverage of Concessional Capital. https://www.convergence.finance/resource/35t8IVft5uYMOGOaQ42qgS/view#. Accessed 10 Aug 2022.
13. Convergence (2021) The State of Blended Finance. https://www.convergence.finance/resource/0bbf487e-d76d-4e84-ba9e-bd6d8cf75ea0/view. Accessed 10 Aug 2022.
14. Development Initiatives (2016) Blended Finance: Understanding Its Potential for Agenda 2030. http://devinit.org/wp-content/uploads/2016/11/Blended-finance-Understanding-its-potential-for-Agenda-2030.pdf. Accessed 10 Aug 2022.
15. Convergence (2021) The State of Blended Finance. https://www.convergence.finance/resource/0bbf487e-d76d-4e84-ba9e-bd6d8cf75ea0/view. Accessed 10 Aug 2022.
16. Learn more by visiting: https://www.macfound.org/programs/catalytic-capital-consortium/.
17. Note that while DFIs deploy the capital the provider is often (but not always) a donor.
18. Interview with Pomona Impact founder Richard Ambrose.
19. Learn more by visiting: https://incofinfaf.com/.
20. Some working at DFIs would argue that they are already being catalytic in their use of blended windows but more work can be done.

12

Scaling Blended Finance Effectively

Despite the catalytic potential of blended finance, its use remains far too subscale. According to the most recent Convergence State of the Sector report on the global network of blended finance, nearly 680 closed blended finance transactions have taken place to date, totaling just over $160 billion from 1450 unique investors, which averages to $9 billion annually.[1] This represents less than 2 percent of annual official development assistance (ODA) by governments and less than half a percent of the annual financing gap required to achieve the SDGs by 2030.

A recent webinar I participated in with the former CEO of the International Finance Corporation (IFC), Philippe Le Houérou, gives hints at why blended finance isn't scaling, despite its potential. On the webinar I asked Philippe why the IFC hadn't fully utilized its blended finance window of donor funding during his tenure as CEO. His twofold response brought forward the main critiques of blended finance. First, he stated IFC needed to define a clear rationale for use of concessional financing so as not to distort markets. Second, he explained that he was nervous about the ineffective use of donor money to catalyze private investors, thus losing future opportunities to raise donor funding to support IFC operations. These are valid concerns, but neither in my view is sufficient to slow the growth and importance of blended finance or the role that DFIs should be playing in its proliferation. DFIs need to take more risks and convene more actors in the scale up of the proliferation of blended finance.[2] In this chapter I will explain why, but first

K. Hornberger, *Scaling Impact*,
https://doi.org/10.1007/978-3-031-22614-4_12

we should look closely at what blended finance is and when and how to use it.

Overview of blended finance

As I noted in the last chapter, blended finance is the strategic use of grants or philanthropic funds, such as those from governments or foundations, to mobilize private capital for social and environmental results.[3] Most often blended finance combines capital with different levels of risk and return expectations. Providers of the more risk-tolerant "catalytic" capital in blended finance structures aim to increase their social and/or environmental impact by accessing larger, more diverse pools of capital from commercial investors or investors seeking a higher financial return.[4] Most providers of catalytic blended finance are motivated by the potential impact that can be generated. On the other hand, most market-return-seeking investors will only participate when return expectations are in line with market expectations based on perceived risk.

To do this in practice, grant or philanthropic funds providers can either de-risk or enhance the returns of an investment to make it attractive for traditional investors along the risk-return continuum (see Fig. 12.1). Most blended finance focuses on de-risking, but it is worth noting that enhancing returns, while less common, is a viable way to attract market-seeking capital using blended finance.

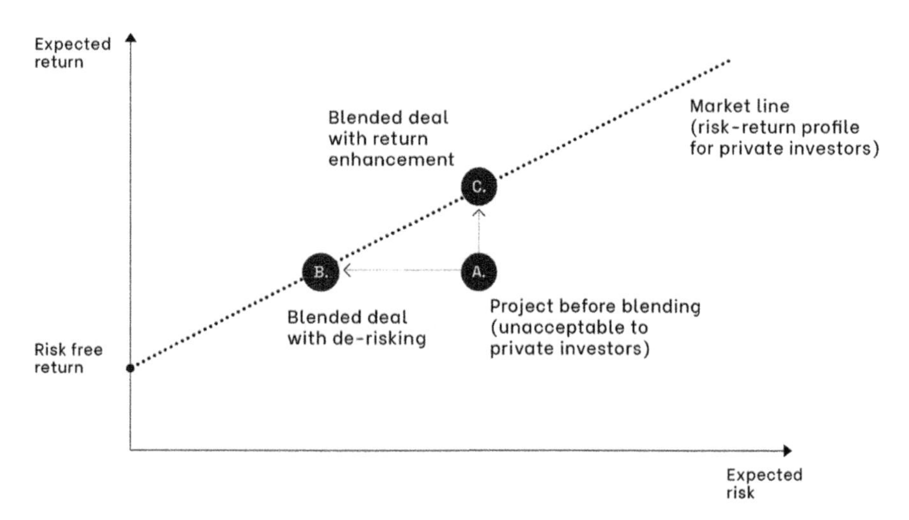

Fig. 12.1 Deploying blended finance to achieve commercially acceptable risk return

Blended finance in practice

The fundamental purpose of blended finance is to mobilize private capital in a way that increases and/or accelerates private capital investment. In this way, the role of the public or philanthropic investor is truly catalytic and does not displace private capital, although this balance is often difficult to strike and the essential challenge of deploying blended finance well. In every instance, the public or philanthropic investor needs to make a clear—ideally evidence-based—case that blended finance is enabling rather than redundant or cannibalistic.

Blended finance cannot replace all development funding, but it can help public or philanthropic investors achieve more with less, particularly in contexts in which market actors can help address development challenges, such as access to essential goods and services. For public or philanthropic investors, the use of blended finance requires some level of underlying market activity and a clear revenue model for the funded activity. This is not the case for most development spending, so determining whether blended finance is appropriate or not requires careful deliberation.

Archetypes of blended finance

One reason blended finance can be misunderstood is because grant providers can de-risk or enhance the returns of traditional investors in several different ways. While organizations such as the World Economic Forum, Convergence, the GIIN, and USAID have slightly different categories of blended finance, I like to group them into five broad archetypes that address both demand and supply side pain points:

Archetype	Description	Types	Pain point addressed
Design funding	The transaction design or preparation is grant funded	• Project preparation facility; • Structuring and Legal advisory	**De-risking**: supply side

(continued)

(continued)

Archetype	Description	Types	Pain point addressed
Guarantee	Public or philanthropic Investors provide guarantees or risk insurance that investments will be paid back if they underpreform expectations	• Investment guarantee; • Portfolio guarantee	**De-risking**: supply ride
Technical assistance	Transactions are associated with grant-funded technical assistance facilities, pre-or post-investment, that support the investment vehicle, or its investment portfolio companies	• Pre-investment pipeline generation and investment facilitation; • Post-investment support to enhance impact and commercial viability	**De-risking**: demand side
First-loss capital	Public or philanthropic investors are concessional within the capital structure	• Junior equity; • Subordinated debt semi-colin • Grant	**Return enhancement**: Supply side
Incentive payment	A donor funds a program that gives direct grants or incentives to investment funds or financial institutions that serve specific segments or meet pre-defined objectives	• Outcome payments; • Origination incentive; • Pay for results	**Return enhancement**: Demand side

Most commonly, design funding, guarantees, and technical assistance are seen as tools to de-risk investments, whereas first-loss capital and incentive payments are means to sweeten the returns for investors higher in the capital stack who may have been unwilling to invest otherwise. These blending archetypes are not necessarily mutually exclusive and in fact are often used in concert. From my own analysis of historical blended finance transactions, nearly 40 percent have multiple blending archetypes (see Fig. 12.2), of which concessional first-loss capital coupled with technical assistance facilities is the most common. For example, I recently supported the design of a blended

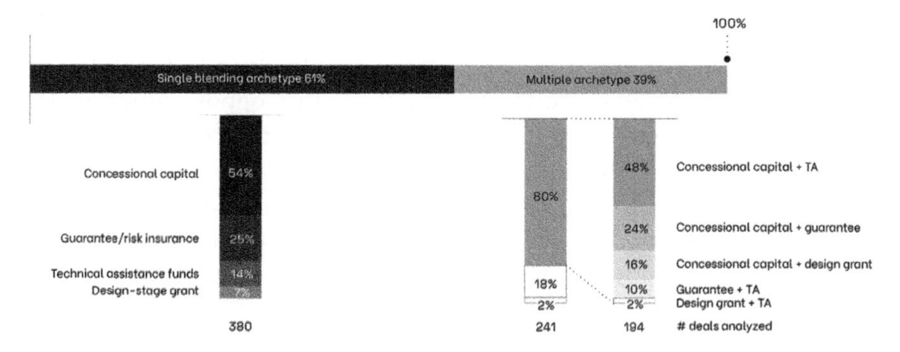

Fig. 12.2 Use of blended finance by archetype (%) (*Source* Convergence historical deals database, accessed May 2021 [653 deals analyzed])

finance fund focused on regenerative agriculture in Central America and included both first-loss capital in the form of subordinated debt and technical assistance to support portfolio companies adopting ESG practices.

Each blending archetype has nuance. Understanding that nuance, in addition to the instrument, is critical to determining when to use each blending archetype. Let's review each archetype in turn:

Design grants and project preparation facilities help investors, banks, and other sources of financing understand market conditions and identify investment opportunities. These commonly occur early in the development of a new market or investment opportunity area. Offsetting structuring and project preparation costs so that the execution appeals to private sector actors and simultaneously offers social/economic development benefits is the primary motivator for using this archetype. Activities include helping structure blended finance vehicles, investment platforms, and other financial products. Financial support also can be provided to cover engineering, consultant, or legal fees.

Design grants are incredibly effective ways to support organizations thinking about using blended finance to scale impactful operations but that lack the experience of doing it and thus need design support to pay for engineers, consultants, and lawyers to help imagine and structure the deals. Project preparation is most frequently used in the infrastructure space where deals are larger and considerable lead time to do engineering and environmental feasibility studies are necessary precursors to making sound investment decisions. I recently had the opportunity to work with the Millennium Challenge Corporation (MCC) and Africa 50 to design a project prep facility called the Millennium Impact Investment Accelerator (MIIA), which aimed to prep deals in social infrastructure (water, sanitation, health, and education services) in Sub-Saharan Africa, which are often underinvested because of deal size and unattractive deal terms. In this case, project prep can help lower the transaction cost and attract larger pools of funding. Design-stage grants are not exclusive to large infrastructure opportunities; in fact, we recently have been seeing more interest in using such grants for smaller blended finance fund design needs.

Another good example of effectively designed staged grants using blended finance is the CrossBoundary Energy (CBE), initiated in 2013, launched in 2020, and expanded in 2022. CBE is a developer, owner, and operator of distributed energy solutions for enterprises across Africa. They used preliminary grant funding from USAID's Power Africa and $2.5 million in design-stage grant funding from Shell Foundation to design the initiative and support distributed energy projects to be developed before deploying capital into them. To date, CBE has developed a portfolio of $188 million in projects for 30 corporate customers across 14 countries in Africa, showing that the initial design-stage funding has paid off.

Guarantees and risk insurance provide credit enhancements, enabling access to resources on better terms. They also cover part of the risk in the event of losses or defaults. Donors can provide guarantees or insurance on below-market terms. These products can also be used to launch a risk mitigation vehicle adapted to a particular market risk (e.g., currency risk, liquidity risk). Guarantees are often more appropriate for more mature market segments and where due diligence to ensure market participants will value the guarantees can be validated. I have seen this to be true with structures de-risking commercial banks in the renewable/off-grid energy sector or providing local currency lending.

Guarantees require a lot of nuanced thinking, and as I have found often, they do not always serve their intended purpose. The logic behind a guarantee is that it will be used to pay back an investor from outside the blended finance vehicle in the event that the vehicle underperforms. Often guarantees are structured to pay back less than one hundred percent of the value to avoid encouraging excessive risk-taking. The challenge I have seen, including with several guarantee facilities set up in Sub-Saharan Africa focused on supporting commercial banks to lend to riskier agricultural/rural populations or SMEs, is that the banks still won't lend to these entities if they are not creditworthy or don't have collateral, even with the guarantee. Thus, before using a guarantee, it is important to be certain that the capital that is being de-risked truly will change its behavior, otherwise it is a less useful tool than others.

One important instance where I have seen guarantees work effectively is with an organization that I am proud to say I helped design. That organization is Aceli Africa, a nonprofit that exists to incentivize incremental lending to agricultural SMEs in Sub-Saharan Africa. Aceli, as it is known by most, was built out of a collective research effort that Dalberg led with the Council for Smallholder Agricultural Finance (CSAF) and paid for by USAID in 2017–2018. We looked at the true economics of lending to agricultural SMEs and discovered that the average loan of $655,000 dollars was not profitable, given a number of challenges faced by farmers in emerging markets, such as crop seasonality, disease, currency risk, and climate change (see Fig. 12.3). These circumstances were made worse because, in the market segments most in need of additional finance in Sub-Saharan Africa, smaller loan sizes and loans to recipients in less common value chains were the least likely to be profitable.

This led to the conclusion that blended finance, and in specific, portfolios' first-loss (a form of risk mitigation or guarantee) and origination incentives could encourage lenders to increase the volume of lending into the space despite the difficult economics. Our hypothesis has been borne out and the model is working thus far: Aceli currently provides 2 to 8 percent first-loss for any qualified financial intermediary that places an agricultural SME loan in Sub-Saharan Africa below a certain size. In its first full year of operation, in 2021, Aceli provided first-loss and/or incentives to 205 loans, mobilizing $26.6 million dollars in lending with average loan size of $130,000 with a private capital leverage of 12x.[5]

Technical assistance (TA) strengthens the commercial viability of a project at pre- or post-investment stages by developing the capabilities of investment vehicle operators and/or building the strategic and operational capacity of the enterprises they serve. Donors may leverage TA to support a priority geography, sector, or market segment. Additionally, donors may provide financial

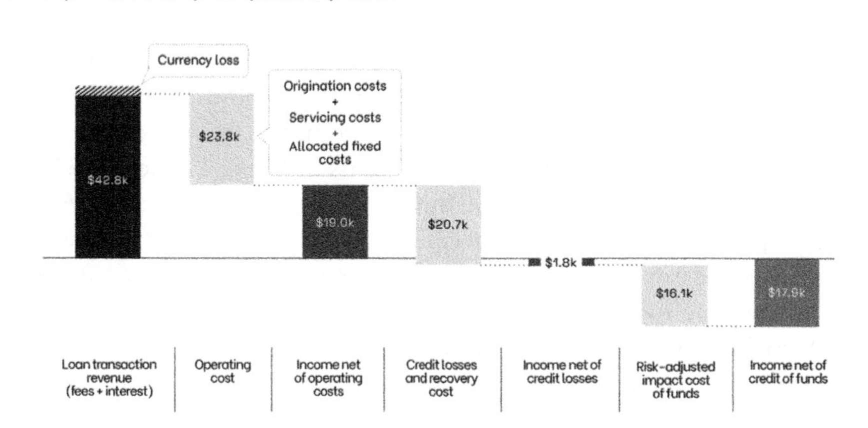

Fig. 12.3 Loan economics averages for all CSAF loans analyzed, $ thousands (*Source* Agrilinks [2018] CSAF Financial Benchmarking Learning Report)

support to offset operational costs of the blended vehicle itself, which can help managers get established and control management fees to attract commercial investors into innovative vehicles. Donors may also provide TA, such as legal and engineering services, to support government entities with private–public partnerships and concessions. TA may also be used to perform transaction advisory services such as investor matchmaking, pitch preparation, financial modeling, and deal structuring. Technical assistance is appropriate in almost all blended finance vehicle scenarios in emerging or frontier markets but continues to be underused.

Technical assistance can be provided pre- or post-investment and can be in the form of providing knowledge or training to the blended finance vehicle organization and staff itself or directly to the enterprises and communities they will be supporting or working within. Technical assistance is particularly useful in scenarios where previous investments have not been made and pipelines need to be developed or matured to the stage they can be investable. Equally useful are enterprise support services that can help enterprises of different types to identify partnerships, suppliers, or markets as well as improve business practices. Finally, technical assistance can also be used to ensure that ESG and impact considerations are taken seriously. Having worked on many technical assistance facilities—including at my first job with TechnoServe Inc. where I helped coffee farmers in Tanzania improve business practices and access markets in collaboration with lenders like Root Capital—I can tell you this is a very underutilized and highly effective blended finance archetype.

Another good example of effective technical assistance facility design was one I came across when evaluating the performance of the lending portfolio of the United States International Development Finance Corporation (DFC). They had provided $4 million in senior debt to an organization called Azure in El Salvador and wanted to understand the impact of the model. What I discovered was that Azure took a very innovative blended finance approach that offered loans combined with technical assistance from Catholic Relief Services (CRS) to water entrepreneurs in Central America. The technical assistance was funded to the tune of $1.2 million by the InterAmerican Development Bank. Since its inception, Azure has funded 17 water service providers and expanded access to clean water and sanitation to 64,000 people in low-income rural and peri-urban areas.[6]

First-loss or concessional capital, on the other hand, improves the risk-return profile for commercial investors by absorbing risk and/or accepting submarket returns for transactions with projected development outcomes. First-loss capital makes concessions that can nurture nascent markets with unproven models, particularly those that serve low-income customers. Donors or philanthropic funders can support investment vehicle operators and projects with concessional financing through subcontracts or grants, such as by providing catalytic capital to fund managers to build first-loss capital into the vehicle structure itself, such as with junior equity or subordinated debt. First-loss capital is best used to attract new or traditional investors that would have been tepid otherwise.

Concessional capital can be placed inside the capital stack of an individual blended finance investment vehicle in many ways. It could be in the form of a repayable grant provided simply to serve as a final backstop in the cashflow waterfall in case the vehicle is unable to make its return expectations and pays back more senior investors first. It could also be in the form of junior equity, which can play the same role but has the advantage of taking some upside of the fund if it is profitable. Or it could be in the form of subordinated debt that is essentially higher-risk debt that can be either secured or unsecured and typically has higher return requirements on interest rate terms to compensate for being paid after senior debt holders in the capital structure.

Determining which one of these concessional capital tools is most appropriate is a dance between the fundraisers and the providers of the concessional financing. I have most often seen subordinated debt being used on the funds I have advised and think it is an excellent tool to encourage other more commercially minded or risk-adverse investors to consider a new sector, geography, or investment strategy or to be more intentionally impact-first in orientation.

Take, for example, the nonprofit fund manager I worked for in the past, Global Partnerships. They have sequentially raised increasingly larger and increasingly more impact-oriented debt funds focused on supporting women and rural entrepreneurs through microfinance-plus models. To get the funding they needed for their $50 million Social Investment Fund (SIF) 5.0 in 2012, they received $8 million in subordinated debt from FMO. They repeated this in 2016 for a $75 million SIF 6.0 with a total of $11 million of subordinated debt in the capital stack from FMO and Rockefeller Foundation. They then raised an impact-first development fund, a $55 million debt fund with $4 million in subordinated debt from Ceniarth and W.J. Kellogg Foundation in 2019. Most recently, in 2021 during the pandemic, they raised a first close for a potentially $100 million impact-first growth fund with $8 million in catalytic funding from Ceniarth and Shelby Cullom Davis Charitable Fund, among others. None of these funds or the lives they have impacted would have been possible without the concessional funding raised.

Incentive payments can be used to encourage providers of capital to serve customer segments that are perceived to be unprofitable or risky. Such payments can be designed in proportion to the size of the investment opportunity and thus encourage activity by sweetening the returns for the investor. They can be effectively used particularly to encourage the extension of existing products into new customer segments. For example, extending agricultural credit to SMEs for new commodities or into new geographies.

When to Use blended finance

To decide when to use blended finance, investors' motivations should be considered. These are varied and driven by a variety of factors such as risk-return preferences, impact targets, and fiduciary responsibilities. Below are examples of situations that often trigger blended finance conversations:

- An investor wants to build a certain size or type of investment vehicle but does not have sufficient capital available through conventional fundraising means without adjusting the risk profile of the vehicle (e.g., a first-time fund manager or investment fund in a new sector/geography).
- An investor has access to grant and/or public money and wants to use this funding to leverage private capital.
- An investor's investment structure is dependent on a combination of public and private capital, such as an impact bond (we will discuss this in Chapter 15).

Whatever the derivation of blended finance, in all cases it must not distort markets, for example by cannibalizing private investment, creating dependencies (e.g., on subsidies), misunderstanding or ignoring local contexts, or neglecting the promotion of high development standards.[7] Lastly, it is critical that the management of blended finance be transparent to ensure that best practice principles are met and to assess the role of blended finance in the context of other development finance and instruments.[8]

The right conditions for pursuing blended finance include:

- Additionality exists (e.g., commercial investors would not otherwise partake or there are other market failures that need to be corrected).
- The impact rationale is clear and ideally linked to the SDGs.
- Insufficient capital is being raised via conventional investment vehicle structures.

- Realistic potential exists to raise grants or philanthropic funding.
- The approach will convey clear benefits to all parties involved in the transaction.
- Organizers share a strong desire for financial innovation and learning by proof of concept.

Determining the right blending archetype(s)

The practice of using blended finance should be flexible and the choice of archetype should be the one that best fits the context and interests of both the investors and the investees involved. Blended finance can take many forms, both within a portfolio of investments as well as within an investment deal or project. Thus, the type of blending archetype most appropriate can vary. For example, in early stages, financing might mostly come from donor-provided catalytic capital, and technical assistance might revolve around foundational groundwork, such as conducting feasibility studies. However, by later stages the financing might be comprised largely of commercial investment, with international development agencies providing a concessional guarantee.[9]

Donors can play a unique role in facilitating transactions by engaging in a set of donor-supported activities that enable private investment, and in some cases, more effective engagement from development finance institutions (DFIs). Donors choose to deploy a specific donor-supported activity or combination of activities based on the level of risk and/or investment return expectations that private sector investors and/or DFIs want to take on in less-developed sectors or markets. Determining the right blending archetype depends on the specific needs of the context in which the blended vehicle is being developed.

Choosing how to structure a blended finance vehicle

Different actors may use a variety of instruments to create various blended finance structures, depending on a variety of factors, including the actors and interests involved, operating and regulatory environments, phase of investment, and existing investment structures, to name a few. That said, designing blended finance well requires a strong understanding of the following elements:

- The appropriate financial instrument or instruments to use
- The underlying model for deploying the instrument(s)
- The structure that deployment takes
- The role of the funders within the structure

Typically, blended finance involves multiple actors along the expected market impact and financial returns continuums (see Fig. 12.4). Though these actors tend to sit at different places along these continuums, it is important to note that blended finance investors may be fluid on the return continuum. For example, though DFIs often channel donor resources as concessional financing to a blended finance structure, they may also act as commercial or senior investors in other transactions. In addition, a single blended finance structure may involve various types of players, for instance, commercial, submarket, and concessional investors all participating together.

Given the diversity of actors involved in the impact investing ecosystem across the risk-return and impact spectrum, great opportunity exists for collaboration to mobilize significant capital toward positive impact. However, the process of designing a specific blended finance approach is not without its challenges. Specifically, traditionally philanthropic and traditionally commercial investors can face difficulties when negotiating on specific terms for investment vehicles as well as lack a common language or shared goals—all of which can lead to costly and longer than necessary design efforts and operationalization of blended finance vehicles. Thus, I strongly advise those considering designing a new blended finance vehicle to engage with experienced advisors who have structured multiple vehicles and understand the interests and motivations of the different parties involved.

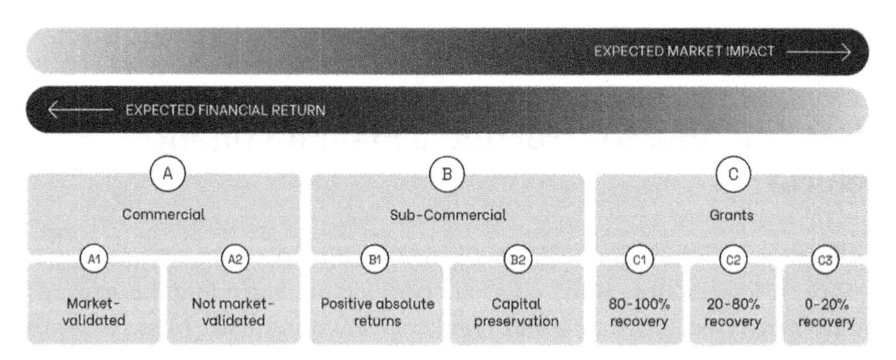

Fig. 12.4 Blended finance investors on the financial and impact returns continuum[10]

Defining the level of concessionality

Concessionality—or what I call subsidy—is embedded within all blended finance archetypes. Since the use of subsidy or concessionality can be market distorting as well as market creating it is incredibly important to define a clear rationale for the use of blended finance. Two primary rationales should be evaluated. First, there must be an economic rationale for the subsidy. Concessional finance should be used for projects that contribute significantly to market development and where the concessional funds are required beyond the public or philanthropic investor's normal additionality to make projects viable.[11] Second, the use of subsidy should minimize concessional funding while maximizing the crowding-in of private capital. This is because concessional funding is scarce and likewise should never cannibalize commercial funding.[12]

To get a sense of both how blended finance is deployed and what concessional levels are appropriate, the IFC shared average concessional levels in a sample of its blended concessional finance portfolio between 2010 and 2020. Overall concessionality rated between 1.5 and 10.1 percent of project value, with an overall average of 3.8 percent.[13] This demonstrated that often only a small share of concessional financing is needed to spark significant investment from sources that would not have otherwise participated. However, from my own experience, far higher levels of concessional financing are often needed to get markets moving, especially for smaller transactions. Development finance actors should be thinking more in the range of 20–30 percent of deal value rather than less than ten percent. For example, several first-time fund managers I have supported investing in emerging markets have relied on first-loss capital, operating support, and technical assistance to cover 30 percent or more of their first-time funds that at $15–20 million in total fund size are difficult to bring to life without blended support.

Understanding financial and nonfinancial additionality

Additionality provides a prime rationale for donor or philanthropic engagement in blended finance opportunities. Additionality means "an intervention will lead, or has led, to outcomes which would not have occurred without the intervention."[14] For example, a project can be considered additional if GHG emissions are lower than they would have been in the absence of the project. Without additionality, providers of concessional funding run the risk of wasting resources on transactions or projects that would have been equally successful without their involvement. Worse, without clear additionality private shareholders may receive funding at the expense of projects in sector or regions where it is most needed.[15]

Considering the relatively scarce availability of grant funding sources as well as the potentially large opportunity cost for using them to crowd in private capital rather than directly fund social or environmental outcomes, there needs to be a strong case for additionality and a link to human or environmental impact to justify its use (see Fig. 12.5). The problem is that most people still think about blended finance additionality only in terms of financial leverage (e.g., how much additional private/commercial funding a blended vehicle attracted that would not have come otherwise). This is important but not the whole story of additionality.

Equally important to financial additionality is nonfinancial additionality, often referred to as development or ecosystem additionality. Nonfinancial additionality can include demonstration effects or proof of concept that is replicated within or across sectors or geographies. It can also include factors even harder to attribute, such as sector-wide growth or the strengthening of the enterprises, funds, and financial intermediaries in previously underdeveloped or overlooked areas, such as fragile and post-conflict countries. Finally, it can be about standard-setting or regulatory policy change where activities lead to improvements in practices related to environmental, social, and governance standards or advances in innovative or customer protection policies.

All of these can be valid reasons to use blended finance beyond the financial leverage it creates. Blended finance creates additionality in situations where commercial financing is not currently available and where an explicit focus is

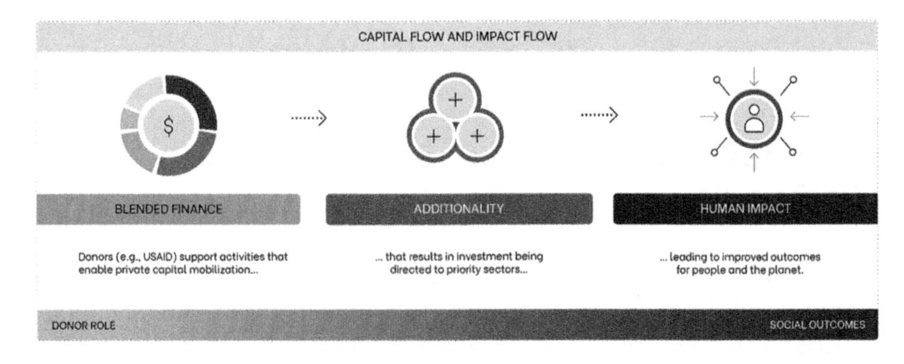

Fig. 12.5 Conceptual framework for blended finance (*Source* USAID [2021], Using Blended Finance to Generate Additionality and Human Impact: Guidance Note)

on opportunities to crowd financing from commercial sources into transactions that deliver development impact.[16] Nonfinancial additionality can often be greater in circumstances where financial leverage is lower.[17] Markets with untested business models, high financial risks (e.g., absence of creditworthy borrowers or high-risk end beneficiaries), and limited track records of private investment may offer higher overall additionality while requiring higher levels of donor support and, initially, lower levels of capital mobilization.

Yet there is no question that DFIs and other actors still face challenges in demonstrating financial and nonfinancial additionality. To do so would require rigorous quantitative evidence based on investment data and credible methods of estimating the counterfactual (what would have happened otherwise)—such as randomized control trials or natural experiments.[18] Neither of which are easy to create on a regular basis and certainly not for DFIs. Thus, alternative means to demonstrate additionality are needed but not easily found. Two promising approaches are the use of auctions and outcomes-based payments (to be discussed in Chapter 15).

Summary of key messages from this chapter

- Blended finance continues to be subscale in its use since there is often a lack of clear understanding of how and when to use it or a lack of clear rationale for why it is a better use than others of a subsidy for grant funding.
- At its core, blended finance either de-risks or improves the return for investors that would not otherwise invest in an investment opportunity based on its risk-return profile. Five archetypes of blending allow grant or philanthropic funding providers to do this. They can provide (1) first-loss capital, such as subordinated debt or junior equity, or even direct grants into the capital stack, or (2) incentive payments to boost return potential for other investors. Or they can provide (3) design-stage funding, (4) guarantees, or (5) technical assistance facility funding to de-risk the investment opportunity.
- The most appropriate type of blending archetype or combination of blending archetypes will vary. In the early stages, financing might mostly come from donor-provided first-loss capital, and the technical assistance might revolve around foundational groundwork, such as conducting feasibility studies. However, by later stages, the financing might be comprised largely of commercial investment, with international development agencies providing a concessional guarantee.

- The use of concessionality via blended finance must have a strong rationale for its financial and nonfinancial additionality. Financial additionality refers to the amount of financing the subsidy can bring to the deal that would not have happened otherwise, whereas nonfinancial additionality refers to the ability to provide demonstration effects, build new markets, or set standards in a new area of business activity. It is also important that subsidy use is the minimum possible to achieve the financial and nonfinancial additionality so as to not distort markets.

Notes

1. Convergence (2021) The State of Blended Finance. https://www.convergence.finance/resource/0bbf487e-d76d-4e84-ba9e-bd6d8cf75ea0/view. Accessed 10 Aug 2022.
2. Gupta S, Walkate H and van Zwieten R (2022) Governments: Blended Finance is Like Music. https://illuminem.com/illuminemvoices/b5f42455-398d-4ee8-b097-bab13e06cc63. Accessed 16 Oct 2022.
3. World Economic Forum (2015) A How-To Guide for Blended Finance. https://www3.weforum.org/docs/WEF_How_To_Guide_Blended_Finance_report_2015.pdf. Accessed 10 Aug 2022.
4. Global Impact Investing Network. Blended Finance Working Group. https://thegiin.org/blended-finance-working-group. Accessed 10 Aug 2022.
5. Aceli Africa (2021) 2021 Learning Report: Year 1: Unlocking Private Capital for African Agriculture. https://aceliafrica.org/aceli-year-1-learning-report/. Accessed 10 Aug 2022.
6. U.S. International Development Finance Corporation (2022) Expanding Access to Clean Water in El Salvador. https://www.dfc.gov/investment-story/expanding-access-clean-water-el-salvador. Accessed 09 Sept 2022.
7. IFC (2021) Using Blended Concessional Finance to Invest in Challenging Markets: Economic Considerations, Transparency, Governance, and Lessons of Experience. https://www.ifc.org/wps/wcm/connect/1decef29-1fe6-43c3-86c7-842d11398859/IFC-BlendedFinanceReport_Feb+2021_web.pdf?MOD=AJPERES&CVID=ntFHkEh. Accessed 10 Aug 2022.
8. IFC (2021) Using Blended Concessional Finance to Invest in Challenging Markets: Economic Considerations, Transparency, Governance, and Lessons of Experience. https://www.ifc.org/wps/wcm/connect/1decef29-1fe6-43c3-86c7-842d11398859/IFC-BlendedFinanceReport_Feb+2021_web.pdf?MOD=AJPERES&CVID=ntFHkEh. Accessed 10 Aug 2022.
9. USAID INVEST (2020) USAID INVEST Blended Finance Starter Kit: 10 Questions about Mobilizing Private Capital for Better Development Results. https://www.usaid.gov/sites/default/files/documents/1865/BlendedFinanceStarterKit1.pdf. Accessed 10 Aug 2022.

10. Adapted from: Bannick M, Goldman P, Kubzansky M et al (2020) Across the Returns Continuum. Omidyar Network. https://omidyar.com/wp-content/uploads/2020/09/Across-the-Returns-Continuum.pdf. Accessed 10 Aug 2022.
11. IFC (2021) Using Blended Concessional Finance to Invest in Challenging Markets: Economic Considerations, Transparency, Governance, and Lessons of Experience. https://www.ifc.org/wps/wcm/connect/1decef29-1fe6-43c3-86c7-842d11398859/IFC-BlendedFinanceReport_Feb+2021_web.pdf?MOD=AJPERES&CVID=ntFHkEh. Accessed 10 Aug 2022.
12. IFC (2021) Using Blended Concessional Finance to Invest in Challenging Markets: Economic Considerations, Transparency, Governance, and Lessons of Experience. https://www.ifc.org/wps/wcm/connect/1decef29-1fe6-43c3-86c7-842d11398859/IFC-BlendedFinanceReport_Feb+2021_web.pdf?MOD=AJPERES&CVID=ntFHkEh. Accessed 10 Aug 2022.
13. IFC (2021) Blended Concessional Finance—The Benefits of Transparency and Access. EM Compass Note. https://www.ifc.org/wps/wcm/connect/290bc660-89bc-4d23-9fd9-2463548fd925/EMCompass-note-105-blended-finance-benefits-of-transparency-and-access.pdf?MOD=AJPERES&CVID=nFDmhmu. Accessed 10 Aug 2022.
14. OECD (2021) Evaluating Financial and Development Additionality in Blended Finance Operations. https://www.oecd.org/dac/evaluating-financial-and-development-additionality-in-blended-finance-operations-a13bf17d-en.htm. Accessed 10 Aug 2022.
15. European Union (2020) The Use of Development Funds for De-risking Private Investment: How Effective Is It in Delivering Development Results? https://www.europarl.europa.eu/RegData/etudes/STUD/2020/603486/EXPO_STU(2020)603486_EN.pdf. Accessed 09 Sept 2022.
16. OECD Principle 2: Design Blended Finance to Increase the Mobilisation of Commercial Finance. https://www.oecd.org/dac/financing-sustainable-development/blended-finance-principles/principle-2/. Accessed 10 Aug 2022.
17. OECD Principle 2: Design Blended Finance to Increase the Mobilisation of Commercial Finance. https://www.oecd.org/dac/financing-sustainable-development/blended-finance-principles/principle-2/. Accessed 10 Aug 2022.
18. Carter, P Wanted: A Mechanism for Additionality. https://carterpaddy.medium.com/wanted-a-mechanism-for-additionality-87fe136e3820. Accessed 09 Sept 2022.

13

"Challenge the Development Finance Status Quo"—An Interview with Joan Larrea

This chapter summarizes my interview with Joan Larrea, CEO of Convergence the leading global network of blended finance practitioners. Joan was hired as Convergence's first CEO and since then has been a pioneer in the education and knowledge building of how to use blended finance for development finance practitioners.

Question 1: What is Convergence, and what is its role in the blended finance ecosystem?

Joan: We operate as the global network for blended finance. We provide data, market intelligence, and deal flow. We operate on a membership basis, but we also do a lot of work that is of public nature. Our role is to accelerate and propagate the use of blended finance as one of many tools—but not *the* tool—to drive more private investment into economies that need to achieve their SDGs.

We operate very neutrally—that is, we do not represent donors or private money. We represent the watering hole everybody comes to when trying to figure out how to use this tool.

Question 2: What is the history and evolution of Convergence?

Joan: Our prehistory goes back to a bunch of smart people from different walks of life walking into a room and thinking about trying to solve several of the reasons why more private sector investment does not go into these economies. Global Development Incubator and Dalberg were there, as was

K. Hornberger, *Scaling Impact*,
https://doi.org/10.1007/978-3-031-22614-4_13

Gates Foundation, the Canadian government, and Citibank. The consensus was that setting up a global center of excellence like Convergence would solve for about four of the seven problems. Issues included finding the money, the money not finding the deals, people not knowing what they are doing, and there being no common language. So we were set up to try to solve some of these development challenges.

We began with a business plan that Dalberg wrote. We started off small, with one donor, which was the Canadian government. We had an arm's-length contract, a contribution agreement with the Canadian government to fund us, and to fund everything we were doing. And we grew from that basis to where we are today, which is that donor contributions are a minority of our revenue line. Our membership is a significant part of our revenues, and we also have a revenue line that comes from fee for service.

Our evolution has brought us to be much more market sensitive because members are not going to pay their fees if we are not delivering value. So, having a revenue line that is a membership revenue line is a way to be market sensitive about whether you are producing value or not. Having contributions is still important because there is an awful lot of stuff we do that is not useful enough for just one party to pay for it. So, evolution-wise, I think our funding base has changed.

Question 3: How do you define blended finance, and what is the rationale for its use?

Joan: We define it a bit differently than you will see OECD defining it for the donor countries of the world, and it is also different from what a Development Bank like IFC would use.

Our definition stems from why we were formed, which is to drive private sector investment—not additional investment, but private sector investment—into the emerging markets. So, our definition is the strategic use of concessional or catalytic capital to drive private sector investment into emerging markets toward the SDGs. There are several different variations of that, but those are the main elements.

It does not encompass every single deal that a Development Bank might do, because our definition of concessional or catalytic capital is not just, "OK, you go out with a long tenor." It is concessional, like you are doing something for which you are not getting paid back for the risk you are taking. When IFC goes out 10 years or DFC goes out 20 years, they are pricing accordingly. That is not quite concessional in our view, otherwise our database would have every IFC deal in it, and that's not really what we're trying to record.

Question 4: What is sufficiently concessional to be considered blended finance?

Joan: What I would say is, any time money is not being compensated for the risk it is taking. So patient debt from Rockefeller, where they are willing to go out long term and not price accordingly, is concessional and therefore blended finance. Patient debt from IFC, where they provide a six-year grace period or permit the borrower to move the principal payment three times but add 300 basis points for this flexibility, is not concessional; rather, it is appropriate pricing. So, for us, there is that test. There is a lot of squishiness around the details, but that is the principle of it.

Question 5: Which of the blending archetypes do you think provides the clearest use rationale?

Joan: We describe four archetypes because they show up so often, but we do not mean for them to be a delimiting list. There are more. For example, affordability gap payments: A water treatment plant is going to be built, and if it were built on commercial terms the water coming out of the plant would be too expensive for a utility, for people to buy it. So somebody steps in and pays five million dollars of the project's cost. That is not one of the archetypes we talk about a lot, but it is real. Or, debt whose interest rate reduces when you hit impact milestones. That is not something described in the four archetypes, but it is real. So there are a lot of archetypes out there. Our four archetypes help us when we are sorting out all deal data, to pick out what goes into our deal database. If anything looks like one of those four it goes into our database; if it is something else, then we look at it more carefully.

The four we talk about more often are:

1. Technical assistance paired with investment activity.
2. Design funding, which is where the grant money is the first on the scene and it develops the concept up to a certain point where other money follows it. So it is sort of a time lapse photography blending.
3. First loss is where you have blending when money is going into the deal as a layer of capital, so it is part of the permanent capital structure. It is a grant, thinly disguised as equity, or debt, with expectations and terms more like a grant than an investment.
4. Guarantee or insurance product, which moves risk around in the transaction. So, it might be a partial guarantee of a debt, or an insurance product

protecting the equity. But it is not in the capital itself, it is unfunded—a contingent obligation to a transaction that is meant to step in if something goes wrong.

These are the four that we see most frequently. But there is output-based aid, affordability gap, and other things that pop up as well. As per which archetype has the strongest rationale for use, this isn't easy to answer.

Which is most effective is a very different question from which is more defensible if you are a donor. Often, the defensible is where interest is. If I am a civil servant, putting grant money out the door that came from tax money, I want to be able to defend the story behind why I made that decision. It can be a healthy thing but also an unfortunate one, because it might make you do traditional programmatic support when really something else was needed.

I do not think any single archetype is the best thing. It's like saying what's better: rice, pasta, or a potato? It depends on what you're cooking. I don't know that there is a superior archetype. Obviously, mathematically, the less money out the door to get the deal done, the better. So, if you were to follow that logic, you end up with a guarantee or an insurance product in which you have infinite money catalyzed per dollar out the door because you do not actually put any money out the door. But I'm not sure that's the right way to think of it, so honestly, I am agnostic on which one is better. It is different horses for different courses.

Question 6: Which of the four archetypes would you say is in shortest supply, or is often the hardest to get?

Joan: Probably the deepest concessional capital, which is something that performs like equity but is not taking equity upside. Equity of course is the most volatile layer in a capital stack. On a balance sheet, anything equity can lose everything or can gain many times its value. And equity is usually the first on the scene in a non-blended transaction. If I want to build a power plant, I first must put up the shareholder money before any lender will talk to me. And it is at that moment that many deals do not happen because the equity providers cannot take that risk, and there is nobody else on the scene. So probably that I would say. Grant capital performs at the very bottom of the capital stack.

There is a pattern among some donors of moving up the seniority toward being a kind of a lender. For example, returnable grants. If you really think what a returnable grant is, it is a loan without an interest rate. They want their money back. Once you are asking for your money back, you are much less catalytic because you are one of many parties in line who want their money back. And it is possible that is the tipping point where other people will not do the transaction because, when they look around, the probability of failure, times the number of parties who want their money back, it is just too much for the transaction to happen—even if that might have actually been a good transaction.

Question 7: How has blended finance and its use evolved since you started working on it?

Joan: At the start, a lot of the interest came from impact firms. They were already working with low-return scenarios, and in some parties getting no return in their transactions, so impact finance people got the point of blending. What was harder was to convince commercial parties that there was a way for them to earn what they normally would want to earn in tougher situations by using some concessional money in their transactions.

So, since I have been at Convergence, I have seen big money parties awaken to the idea that maybe there is a way to work with donors to do things they want to do. If commercial investors are not in certain markets or certain situations, or have turned things down because they do not quite have the return profile, now maybe they can do them. Such investors are not everywhere. For example, they are not wanting to do early childhood development. Rather, they want to do climate finance, and particularly the renewable side of climate finance, because they think they understand that a little bit better.

I think that bigger, more institutional actors have stepped forward. And that is exciting. I have not yet seen the donor response of "Let's serve up some things you can invest in"—we are not there yet. So that has changed. We are gradually seeing those who want to get large transactions done considering blended finance.

Question 8: What are some examples of the results that have been achieved through the use of blended finance?

Joan: Unfortunately, the results and impact are hardest to capture, especially for us as a network association, since we are not principals in deals—few organizations report to us on a voluntary basis. So, capturing the ex-post is really hard for us as a network. That said, I think one result of blended finance is accelerating progress—accelerating transactions to close so that they can get started. For example, we provided a grant in our design funding program to Nature Conservancy, a debt swap for nature. They shopped the deal around for a long time, so by us stepping in, helping design the deal, and signaling our support, we were able to help speed up the process. And I do think just acceleration of getting things to market is an outcome of blended finance that is legitimate and important.

Another one is scale. There have been a lot of deals that have happened in an incremental manner that built up scale over time. One example is Climate Investor One, a set of three funds. The first put together large amounts of grant funding focused on the development stage when renewable energy developers cannot get traction. The second fund was blended and focused more on construction finance. Finally, the third fund, unblended, focused on holding operating assets. That whole idea allows a 1–2–3 support system for renewable power developers, so project developers are not doing one painful thing and then going to the next stage doing another painful investment raise with another funder thing. They have a place to go where they can get an A-C solution. Blending sometimes has the effect of hitting scale, whereas before you were doing things without support.

The third is correcting for market imperfections. And we all know that while markets are dynamic—pricing follows demand and supply—they are also imperfect. People's information set is imperfect. Their sense of risk is imperfect. And sometimes you can have a wonderful deal, but it is so small that it cannot get financing. There is that imperfection. So, I think sometimes blending cures market imperfections.

If you think about one massive example, it is all of the advanced market commitments in the healthcare industry, where Gates or whoever stood up and said, "We promise if you make this vaccine, we will buy at least X million doses of it." And the reason that the vaccine was not being produced was because the makers did not know if anyone would buy it, because the people who are buying it are poor countries and NGOs, and who knows if they will ever pay for it. So you can cure a market imperfection with blended finance.

I think these are some of the effects of blended finance: speed, scale, and addressing market defects.

Question 9: What do you think about the idea of ecosystem additionality?

Joan: I strongly believe in ecosystem additionality. I think doing one-off transactions that work is not the point. There needs to be two kinds of off-ramps. Obviously, in any deal, the donor should be asking, "Can I get out of this, and have this deal stand on its own over time?" But also, each transaction should be laying down a marker that says, "This kind of deal is doable," so that somebody else will follow without blending in deal number two, three, and four.

I think ecosystem additionality also takes the form of parties who never talk to each other, talking to each other. USAID will never naturally talk to a JP Morgan Chase. But if they talk to each other, they will learn something. When you do a blended finance deal, somebody is there not for the return, but for the impact. So, they must have something to teach the parties that are there for financial return. And those who are there for financial return think about every single dollar and what it is going to bring back. They do not spend one cent unless they know where it is coming back from. That discipline and long-term perspective—for example if you are a private equity investor, you are thinking "How am I going to make this company four times its size in ten years." In addition there is the additional benefit of learning from others in the ecosystem people do when they form blended finance structures.

Question 10: What are the biggest challenges to the scale and use of blended finance? Particularly for development banks?

Joan: First of all, the obvious observation is that blended finance takes a source of concessional capital. That is the magic money, the one that is in short supply. It is in short supply because, of course, of the budgets of countries and philanthropies are constrained. But also, it is in short supply because there is no strategic commitment by donors to support private sector activity as one of their impacts. There is no donor out there that will say, "It is part of my mission to get private sector money into my countries, and I will dedicate 10 percent of my resources to it." Nobody says that, so there is a supply problem on the donor side and that contributes to the problem of lack of scale. Obviously, if we use what resources there is in a much better way, we can defeat part of that, but that is still a large constraint.

There are several other challenges, in no particular order:

Lack of standardization. Everybody wants to have very crafted, specific deals that work just in this one instance and work beautifully. And everybody thought of every angle, and there is minimum concessionality. How do you test for that? What is the exact right number of dollars of grant money for this specific deal? I do not know. Do you want to debate it or get on with it? We will get to scale when people have a standardized approach.

Generally, in a private equity fund, you do not want to see more than X percent be from the donors, unless you are in a frontier economy, in which case it will be slightly higher at Y percent. That kind of standardization needs to come out so that we don't spend three years designing these things. If you think about how project finance works, where everybody knows there is an offtake, everybody knows there is an engineering, procurement, and construction (EPC) contract, everybody knows there is a performance bond, and everybody skips to page 29 of the legal documents because they know what pages 1–28 are. We need to be there, and we are not. So, I think blended finance needs to come out to that moment, like project finance did.

Other things not there yet, namely domestic actors and domestic capital. That is a problem for development finance generally, but can we somehow make better use of blended finance to pull out domestic capital markets and domestic players?

And there we come to the multilateral development banks (MDBs). A large chunk of blended finance flows through development banks. A donor will give a trust fund to a development bank. The development bank will blend and put a transaction together with that blended source. If that is the case, then we need them to be thinking about blending in a different way. Right now, as far as I can tell, the blending is to reduce the risk of the transaction for the development bank. There is sort of a bifurcation of thinking. For the transactions that are hard to do, we will go get donor money, and we will do them. Separately, for the transactions that you can syndicate, because they are in a medium-income country and they are in a known sector with lots of history, we will do a syndication and we will mobilize other investors to invest alongside us, or through our syndication platform.

Why not do more thinking around using donor resources as a multilateral to design a transaction that other people can step into with you, so you are US$1 out of US$10, instead of all the money on the table. Be the designer of the transaction, the facilitator of the transaction, and a participant in the transaction. And do much more to pull in other people's money. We are not there, and it is not because the development bank personnel do not know what they are doing. They are good and talented, and they deliver exactly what stakeholders ask of them, and the shareholders are not asking for massive rethinking of mobilization. And when they do that, if they tell development banks, "You are supposed to be mobilizing many times what you are doing today," then all resources will be put on it, including probably donor money that they have sitting in trust funds.

Question 11: What is next for Convergence? How do you see the community continuing to evolve over the next three to five years?

Joan: To start, I think the institutional investors coming in is a really exciting development, and we are all over that. Yes, some of it might be talk, some might be impact washing or greenwashing, but that is a first step, and we at Convergence need to respond to that. We are working really hard to get those sorts of parties into our membership. The Net Zero Asset Owners Alliance, for example, has a sort of a composite membership with us. And we are trying to speak to them and not just to donors. When we think about what we should write next, when we should convene next, and what we should talk about next, we are also trying to serve that constituency. The whole institutional investors showing up at the party is a pretty big deal for us, and we need to respond. We hope that donors respond as well, but that is big for us.

I think we are switching from a neutral observer tone all the time to a more persuasive voice. When we started, really the field had not been documented, so we spent a lot of time explaining to people what it looked like and making observations, pattern recognition, etc. Now I think we have a firm basis on which to say, "This is really what it should look like." And again, we are not representing any constituency; it is just if this tool is going to be useful, these are some of the things everybody needs to be doing. I do not think we'll ever turn into a lobbying or advocacy organization; we do not have the skill set. But we can be just sort of an honest broker, an opinionated voice.

We are also working on the impact narrative. We need to have a strong response to those who say, "It is all about subsidizing the larger commercial investors." Part of the answer is, Why do you care if somebody gets richer if you get more impact units per dollar? You actually want them to get wealthy, right? Especially if it is a domestic business. They took a huge risk and you helped them take their risk; don't you want them to thrive and hire 20,000 more people? But no, sometimes culturally that is a hard sell.

Part V

Measure Success Based on Results, Not Activities

14

The Unrealized Potential of Results-Based Finance

As one of my first assignments after I joined Global Partnerships—an impact-first nonprofit fund manager based in Seattle was to evaluate a potential investment in Clínicas del Azúcar (CdA), a health company focusing on diabetes treatment in Mexico. I quickly got to know Miguel Garza and Javier Lozano, the co-founders of CdA in Monterey, who had a vision of scaling up low-cost and comprehensive diabetes care for the large and growing population that suffered from the condition in Mexico. Through initial conversations I discovered that their first clinic was a success but that they were looking for additional funding to expand the number of clinics. Unfortunately, at that point in time their investment needs, profitability, and track record would not allow us at Global Partnerships to use our debt funds to provide them financing. Nor were we certain that debt financing was actually what CdA needed at that point.

However, working in partnership with the Linked Foundation and later Roots of Impact and the Swiss Development Corporation (SDC) we were able to come up with a novel solution.[1] Given CdA was not yet breakeven and subscale, their most important task was to fully embrace its mission and successfully reach patients at the bottom of the pyramid to provide them with needed diabetes care. Thus, a proposal was developed to use a novel results-based financing approach called Social Impact Incentives (SIINC). The SIINC would be structured to provide additional financing to CdA for real impact achieved on diabetes with lowest income populations in Mexico.

K. Hornberger, *Scaling Impact*,
https://doi.org/10.1007/978-3-031-22614-4_14

In specific, additional concessional debt and grant funding would be provided to expand the number of clinics based on two metrics of performance: the ratio of base-of-the-pyramid clients to overall active members (a measure of inclusion) and a measure of improvement in HbA1C levels in those patients (a measure of quality of care).[2]

As a result of performance in the first year, CdA was able to secure funding from Linked Foundation and SDC to expand the number of clinics it operated from one to five. The results continued to be delivered and thus the results-based funding did too. Once CdA reached sufficient scale Global Partnerships (and other private investors) were able to invest and help it to grow further. In fact as of the time of writing CdA has more than 25 clinics across Mexico serving thousands of customers. Through an independent evaluation it was demonstrated that more than a third of its clients were base of the pyramid or the lowest income quintile in Mexico and they have seen an average improvement in HbA1c levels ranged between 2.18 and 2.8 per six-month period, meaning substantial reductions in the risk of diabetes complications.[3] Furthermore, each dollar of SIINC donor funding has leveraged three dollars of private capital investments to fuel its growth.[4]

This is an excellent example of the potential for results-based finance, yet it remains widely underutilized. Overwhelmingly, donor and philanthropic funding for international development continues to focus on activities and outputs rather than results. In 2017 more than 93 percent of all official development assistance by donor governments was output based.[5] In line with funding cycles, many donors continue to favor short-term projects that focus on predetermined outputs. Similarly, year on year, less than 1 percent of foundation giving goes toward results-focused innovative finance projects.[6] Instead, foundations too often allocate resources based on fixed annual endowment contributions that leave program officers clambering to design grant projects to fit the needs of the budget rather than those of the communities or the results they are trying to achieve.

The failings of business-as-usual donor and philanthropic spending practices when tackling complex social problems are widely recognized. Many academic studies have highlighted problems such as the rigidities of traditional programming, disincentives for using evidence, too little focus on results, and an overreliance on "cookie-cutter" approaches rather than searching for context-specific solutions.[7,8]

Having spent nearly my entire career working on social impact projects in emerging or frontier markets, often with donor or philanthropic support, I can confirm this is the sad reality. Nearly all the project proposals I work on at Dalberg, whether for donors, foundations or other providers of grant funding, focus on the activities that will be executed against an often-predefined budget rather than speaking to the ultimate impact we hope to achieve and the necessary outcomes that will drive us there. At best we too often pay lip service with poetic words to the outcomes we hope will come true and rarely sign up to do projects that focus more on what is achieved and less about how we ago about it. It is true that the nature of our work doesn't allow for the time to see the impacts and outcomes of our work materialize. Nevertheless, we spend far too much time working on activities or outputs rather than on the broader, observable effects of the interventions and whether the outcomes we witness can be attributed to the activities being completed.[9]

As long as we keep monitoring and measuring our success based on the activities delivered, there is no incentive to maximize for impact. Unless we zoom in on the linkages between outcomes and activities delivered, we won't truly grasp what makes a project successful and harvest the learnings that can bring them to scale.

Charting a new path forward

For anyone like me who cares about solving some of the world's most challenging problems, such as poverty or climate change, it is critical to focus on improving not just the quantum of capital we put into the system but also the effectiveness of how we spend it. More donors and providers of philanthropy should move away from siloed, rigid, outputs-focused indicators and toward more flexible reporting requirements and longer-term funding cycles. This is a necessary condition for addressing the complexities of systems change required to tackle problems such as gender and racial inequities.

Results-based financing (RBF)—also known as Pay-for-Success (PFS) or more broadly Innovative Finance—is a critical ingredient for driving greater effectiveness of donor and philanthropic spending moving forward.[10] By tying the funding to results rather than to activities and outputs, well-designed RBF introduces performance incentives for funding recipients. Further, it provides implementers with greater flexibility to adjust their programs as needed to changing circumstances, empowering them to innovate, learn, and adapt their programs in pursuit of outcomes and impact.

By embracing RBF, program implementers are also better placed to respond to donors' and foundations' growing desires to demonstrate measurable outcomes from their funding. In addition to demanding more rigorous program evaluations (often with randomized control trials),[11] funders are increasingly demanding that their money be spent in ways that allow them to directly pay for outcomes. Responding to these demands is critical for implementers to truly realize their impact potential, especially in the face of an increasingly competitive funding landscape.[12]

More importantly, since donor governments or foundations only pay the full cost of programs when results are achieved, RBF enables implementers to offer greater value for money. This may help implementers engage funders and secure support for innovative programs—even if many funders are relatively risk averse and if programs are based on nascent evidence or uncertain likelihood of achieving desired results.

The rise of results-based financing

The good news is that use of RBF approaches continues to grow. According to a study done by Instiglio, the value of RBF grew by more than 30 percent per year from 2010 to 2017 to reach $42 billion dollars in total commitments across 51 different projects sponsored by bilateral donors and development finance institutions.[13] This includes programs like the World Bank's Program-for-Results and United Kingdom's Girls' Education Challenge Fund. These programs still represent a small share of overall development assistance funding, but they are not insignificant.

Among all RBF approaches, the use of impact bonds in particular has grown dramatically over the last decade. Using data from Oxford University Government Outcomes Lab, we see that the number of Development Impact Bond projects has grown by more than 40 percent per year from 2010 to reach 33 in 2020 across a wide range of themes such as employment, health, and education (see Fig. 14.1). Some of the notable examples include the $9 million dollar Refugee Impact Bond focused on building resilience for Syrian refugees in Jordan and the International Red Cross Program for Humanitarian Impact Investment of $25 million focused on building and improving staff efficiency at three physical rehabilitation centers in Mali, Nigeria, and the Democratic Republic of Congo. At the same time we have seen a trend reversal from 2017 to 2019 and the use particularly of Development Impact Bonds (DIBs) has slowed down as more funders have become aware of the challenges they present (discussed in more detail later).

Despite this growth in usage, RBF approaches continue to remain far too niche and subscale to make a meaningful impact on the achievement of the SDGs. One of the largest barriers is a lack of understanding about the types of RBF and how and when to use them.

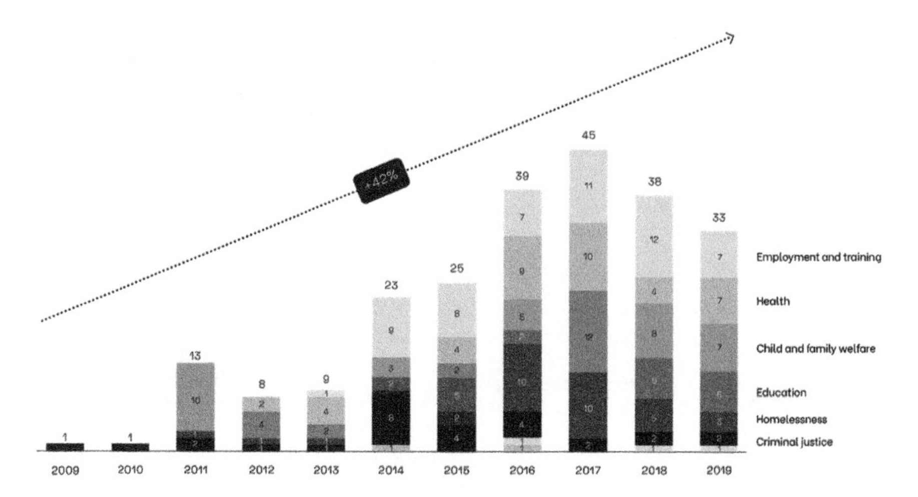

Fig. 14.1 Number of Development Impact Bond projects started by year and thematic focus (*Source* Oxford Outcomes Lab Impact Bonds Dataset, Author analysis)

A menu of results-based financing options

Results-based financing is an aid modality that aims to link payments directly to results, focuses financing more on outputs and outcomes, rather than inputs and processes, with the goal of creating incentives to improve program effectiveness. Results-based mechanisms typically define successful outcomes in advance, and funding is only released upon achievement of these results. This allows for efficiencies in how donor or foundation funding is used and improves the effectiveness of aid funding, while also allowing service providers more flexibility and encouraging innovation in how they carry out their programs to achieve targeting results.

Figure 14.2 illustrates comparisons of a non-exhaustive list of RBF mechanisms based on relative ease of structuring and track records of success in practice. Below I provide short descriptions of these RBF mechanisms.

Impact Bond: Social and Developmental

Impact bonds are results-based contracts in which one or more investors provide working capital for social programs, service providers (e.g., implementors) implement the program, and one or more outcome funders pay back the investors their principal plus a return if, and only if, these programs succeed in delivering results. In a social impact bond (SIB), the outcome payer is typically the government in high or upper-middle-income countries, while in a development impact bond (DIB), the outcome payer is typically

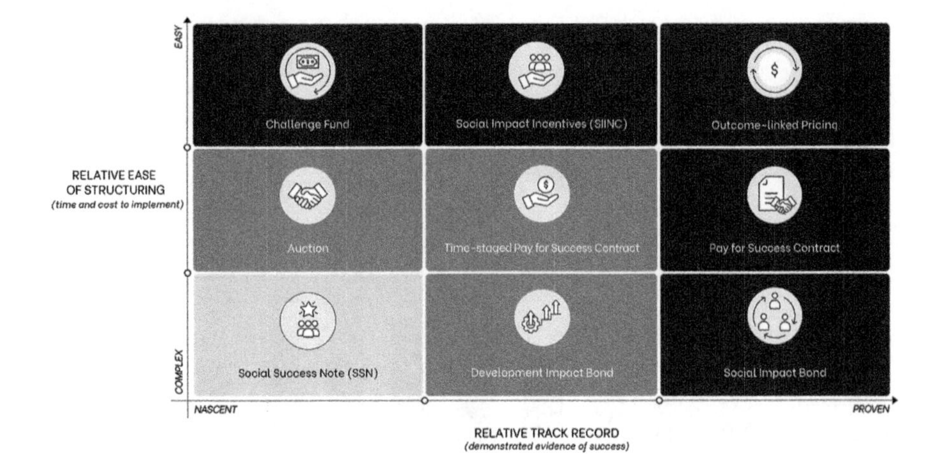

Fig. 14.2 Comparisons of results-based financing mechanisms based on their relative ease of structuring and their track records of success in practice

a private donor or aid agency. The management of an impact bond may be outsourced to an expert bond program manager. In addition, an external evaluator that measures performance needs to be contracted and expenses covered by the donor or philanthropic funding provider.

Impact bonds are most appropriate in scenarios in which an actor is willing to pay for outcomes, but not willing to bear the risk that the activities won't yield the desired results. Impact bonds work best when the knowledge of solution or whether programming will work is unclear and thus outcomes or results risk not being achieved. The primary benefits of impact bonds are focusing on outcomes, building a culture of monitoring and evaluation, driving performance management, and fostering collaboration between stakeholders.[14]

But impact bonds are not the best fit for every context. A donor would generally pay in full for the outcomes, meaning donors have limited leverage to withhold funds as impact bonds permit. Impact bonds are also generally untested for private sector interventions in emerging markets and are less useful when it is known that the solution will achieve agreed-upon outcomes/results, as they have high transaction costs. The life of impact bonds is usually three to seven years, from starting the feasibility study to achieving actual project/program completion with measured results. The cost to structure an impact bond can be 10 percent or more[15] of the total programming cost, and outcome funders will be required to pay investors a premium/interest on top of program costs if (but only if) results are achieved.

As previously mentioned, impact bonds are the most popular of all RBF mechanisms. Relatedly, many people equate RBF with only impact bonds, despite the fact there are numerous of other RBF mechanisms, as illustrated in Fig. 14.2 and described below. Given the emphasis on impact bonds in the RBF world, I dig further into their mechanics and use cases in the next chapter.

Social Success Note (SSN)

A SSN is a variation of an impact bond. It also "pays for success" and crowds in commercial capital to finance social businesses. As with impact bonds, service providers receive capital from an investor upfront in the form of a loan, but instead of requiring the outcome funder (i.e., a donor government or foundation) to repay the investor for the loan provided, repayment is split between the outcome funder and the borrower. The borrower must pay back the principal of the loan with no interest and independent of the results obtained. In addition, the outcome funder pays a premium to the investor, but only if results are met. An example of an SSN is the UBS Optimus and Rockefeller Foundation-funded Impact Water Social Success Note which aims to sell and install 3,600 water filtration systems in Ugandan schools.[16]

SSNs are a useful alternative to impact bonds when there is early evidence that programs will succeed at achieving outcomes or results, but it is not yet fully proven or measured. They typically last from one to five years, including structuring, implementation, and results measurement. The greatest benefit of SSNs is that they better align the incentives of implementing partners or service providers with the outcome funders' interests. However, unlike Impact Bonds, they require that service providers repay the principal of their loan on their own and without investor support. This means that only projects or programs that could be self-sufficient on their own (such as fee for service models) work, and other more traditional development aid programs (such as providing infrastructure) are not feasible in this form.

Outcomes-based contracts

Outcomes-based contracts are bilateral agreements between a payor and service providers. Under the arrangement, service providers receive some funding from the payor to operate the program and receive reimbursement for full project/program costs and/or additional performance payments if they achieve agreed-upon outcomes. An example of an outcome-based contract is the Amazon Fund, a REDD+ mechanism created to raise donations for non-reimbursable investments in efforts to prevent, monitor, and combat deforestation, as well as to promote the preservation and sustainable use in the Brazilian Amazon originally set up by the Brazilian development bank BNDEs.[17]

Bilateral outcomes-based contracts between outcome funders and service providers are best suited to situations where a high degree of confidence exists that the activities within the program/project funded by donor will achieve expected outcomes/results. They are typically short-term (one or two years), simple to structure, and highly replicable. The main drawbacks relate to lack of risk sharing or lack of involvement of third-party investor/working capital providers and limited innovation in delivery, since service providers would likely only sign contracts when they have a high degree of certainty they would be repaid.

Time-staged outcomes-based contracts are a variation of outcomes contracts in which repayment is staged using periodic milestones, such as quarterly or annually. This has the additional benefit of allowing smaller amounts of initial funding.

Outcomes-linked pricing

Outcome-linked pricing are bilateral agreements between an investor and service providers. Under the arrangement, service providers receive working capital from the investor to implement a project, and the pricing of interest rate for repayments is lower or higher depending on achieving the outcomes agreed-upon during project implementation.

Like more traditional bilateral outcome-based contracts, this RBF is most useful when the interventions show evidence that they will work and for shorter-term interventions and results measurement periods. These agreements are simple to structure and are highly replicable. They also align the incentives of service providers with those of outcome funders. However, they typically are not used for programs that require a high degree of innovation over unknown outcomes or that do not provide risk sharing. They are also not useful for programs or projects that do not have the ability to pay back loans, as they are a pricing mechanism on a loan and not a form of grant funding.

Social Impact Incentives (SIINC)

This funding instrument rewards high-impact enterprises with premium payments for achieving social impact. The additional revenues enable enterprises to improve profitability and attract investment to scale. Thus, SIINC can effectively leverage public or philanthropic funds to catalyze private investment in underserved markets with high potential for social impact. An example of this RBF mechanism is Root Capital's Pay for Impact Partnership in Latin America, funded by the InterAmerican Development Bank Lab (IDB Lab) and the Swiss Agency for Development and Cooperation (SDC).

SIINC are a form of incentive and are best used for projects or programs where some evidence of early success has already been demonstrated. They are usually implemented in two- to three-year time frames as contracts between grant provider and service provider. The main benefit is that they do not require full programing costs and only pay for the upside or performance of enterprises/programs/projects. They are also very easy to structure, replicate, and scale. The main drawback is that they can only be used on successful projects and typically require much smaller amounts of funding than do impact bonds or bilateral outcome-based contracts.

Challenge funds

These funds galvanize people outside the funding organization to develop innovative solutions to development challenges. The funding organization awards the prize funds to the organization(s) with the best solution that achieves desired outcomes and helps surface new types of partnerships. An example of a challenge fund is USAID's MujerProspera (WomanProsper) Challenge. The regional challenge is to advance gender quality in El Salvador, Guatemala, and Honduras and awards funding to private organizations that have developed proven solutions to promote women's agency, safety and access to power, resources, fair and stable income, or jobs.

Awards and prizes

Awards or prizes, like challenge funds, are competitions that help surface solutions, partners, and ideas to help support innovation and implementation. Unlike challenge funds, however, awards do not set a condition or a goal, but rather award the prize to the best solution surfaced. Thus, the main drawback of these models is that they may not actually surface good ideas but simply reward the best of the bad ideas/projects that emerge. A great example of this approach is the Massachusetts Institute of Technology (MIT) Solve initiative. Solve is an open innovation challenge that finds tech-based social entrepreneurs, runs an open competition which selects the most promising innovations, and matches them with funding and support from the MIT ecosystem.

Challenge funds or awards are most useful when little evidence or knowledge shows a project or program can achieve the desired outcome or results. Challenge funds and awards can encourage ideation and innovation on hard or difficult to solve problems and can be implemented over defined periods of time. The main drawback is that they can be difficult to replicate and scale due to the context-specific nature of the problems they address and the amount of funding required to make them work.

Auctions

In an auction, funders commit to funding upfront, provide initial funding to implementing partners to get started or pilot, and then award full funding or a share of funding via auction to the most competitive bidder, based on outcomes achieved and cost efficiency. An example of an auction is the World Bank's Pilot Auction Facility for Methane and Climate Change Mitigation (PAF). PAF is an innovative, pay-for-performance mechanism developed to stimulate investment in projects that reduce greenhouse gas emissions while maximizing the impact of public funds and leveraging private sector financing.[18]

Like challenge funds or awards, auctions are useful when knowledge of solutions to an existing problem is scanty. They incentivize innovation and cost relatively little to set up. They also provide the additional benefit of not requiring full funding if no bidders are successful, as the Auctioneer has the option to choose the best solution based on its own criteria. Auctions are typically narrow in focus, however, and difficult to scale.

Mainstreaming results-based financing

Now we turn to the question of how to use and scale these financing options in practice. In the next chapter, I look at how and when to use one of the most popular forms of RBF: the impact bond. Before turning to that, however, let's discuss ways to begin mainstreaming RBF approaches and pivoting donor and philanthropic funding toward results, not activities.

Mainstreaming RBF will require changes to donor government and foundation policies, practices, and systems as well as recruiting and training staff. For example, the widely perceived lack of capacity or knowledge among procurement staff will limit their ability to use and deploy RBF approaches as part of contracting processes. Similarly, barriers such as indicator design and the high costs of monitoring and evaluation will make RBF more difficult to implement but will also have implications for traditional approaches. As such, addressing the barriers to RBF implementation will require rethinking donor and philanthropic operations at all levels.

Unfortunately, the evidence required to support mainstreaming RBF approaches continues to be weak. Despite wider use of RBF, evidence on the performance of RBF, and more specifically of impact bonds, remains nascent. Arguments for and against RBF remain theoretical. Too few examples have been completed, and it is too soon to draw conclusions about efficacy. Thus, donors and philanthropic organizations interested in its potential should focus on accelerating its use and on building the learning and evidence base for when and how to use RBF effectively.

Specific donors and foundations can support pioneering efforts to implement RBFs with training, guidance, and tools. Program offices at foundations and inside donor organizations continue to show great interest in the potential of RBFs, but they don't fully understand how to use them effectively—such as the right process to verify outcomes or how to write contracts. I saw this firsthand through my work advising the Women's Entrepreneurial Finance (We-Fi) Secretariat at the World Bank, where I advised the team on incorporating RBF structures into their call for proposals from DFI members. By promoting knowledge sharing, such as by creating communities of practice that support informal sharing, preparing case studies that highlight success stories, and documenting best practices on measuring results, contracting, and structuring RBFs, it becomes possible to incorporate them into programming decisions.

In parallel, donors and foundations can help to further build the evidence base for RBF by continuing to fund and pilot new approaches or methods for leveraging RBF in practice on new topics and by using new instruments for new desired outcomes and in new geographies. Piloting new RBF approaches should always start with a specific development challenge and an explicit hypothesis as to why RBF is better suited to address that challenge compared to traditional solutions.

Finally, donors and philanthropic funders are needed that will continue to fund rigorous evaluations that assess if widespread RBF would be more efficient and effective. Evaluations should critically assess whether RBF is a better alternative to traditional approaches to funding development, as current research tends to focus more on the impact of the intervention than the link between the impact and the financing model used to fund it. In addition, better approaches to measuring outcomes are critical for RBF. Increased investment in the design, selection, and monitoring of indicators is needed, as well as methodologies for listening to beneficiary and community voices. In particular, reduced costs for mobile and digital technology offer new approaches for monitoring and evaluation that make it more cost-effective than ever to measure the results of programs relying on RBF (see the interview with Sasha Dichter to see how this is being done in practice). Donors and foundations should seize this opportunity.

None of this experimentation and learning with RBFs should be done with the approach of a "hammer looking for a nail." That is, RBF cannot be forced into all contexts. Through a sour experience in Haiti, I learned the hard way that impact bonds don't always make sense; sometimes traditional donor funding is all that is needed (more on this in the next chapter). This caution also means one should be willing to accept a lot of failure and to learn through experience. Sometimes donor resources will be wasted, with only the learning to show for it. At the end of the day, RBF will not be a panacea for solving the SDGs, but it does have potential to reorient the conversation in international development away from activities and outputs and toward results and outcomes. This shift is sorely needed and should be furthered.

Summary of key messages from this chapter

- The major part of donor and philanthropic funding continues to be provided in a way that focuses on activities and outputs and *not* outcomes and impacts.
- Results-based financing mechanisms present an alternative pathway that can reorient grant making toward ensuring the focus is more on results and less on process.
- The many types of results-based finance include the impact bond, social impact incentives, and outcome-based contracts. The appropriateness and use of each contingency-based contracting mechanism is highly dependent on the context of the specific intervention.
- To see results-based financing approaches mainstream will require increased commitment from donors and philanthropic funders to experiment, learn, and share evidence with a broader community. We are still at the early stages of seeing RBF approaches reach their potential.

Notes

1. Linked Foundation, Clinicas del Azucar. https://linkedfoundation.org/project/clinicas-azucar/. Accessed 10 Oct 2022.
2. Roots of Impact (2021) Empowering Clinicas del Azucar to Attract Investment and Create Impact at Scale.
3. Ibid.
4. Swiss Agency for Development and Cooperation (2022) Social Impact Incentives (SIINC): Rewarding for Social Outcomes and Mobilizing Capital for Impact.
5. Author's analysis based on data in: World Bank (2018) A Guidebook for Effective Results-Based Financing Strategies. https://documents.worldbank.org/en/publication/documents-reports/documentdetail/265691542095967793/a-guide-for-effective-results-based-financing-strategies. Accessed 14 Aug 2022.
6. Author's analysis based on data in: Brookings Institution (2022) Global Impact Bonds Database Snapshot. https://www.brookings.edu/research/social-and-development-impact-bonds-by-the-numbers/. Accessed 14 Aug 2022.
7. Rodrik D (2007) *One Economics, Many Recipes*. Princeton and Oxford: Princeton University Press.
8. Bain K (2016) *Doing Development Differently at the World Bank: Updating the Plumbing to Fit the Architecture*. Overseas Development Institute. https://odi.org/en/publications/doing-development-differently-at-the-world-bank-updating-the-plumbing-to-fit-the-architecture/. Accessed 14 Aug 2022.
9. Stannard-Stockton S (2010) Getting Results: Outputs, Outcomes and Impact. *Stanford Social Innovation Review*. https://ssir.org/articles/entry/getting_results_outputs_outcomes_impact. Accessed 14 Aug 2022.
10. Birdsall N, Savedoff W (2010) *Cash on Delivery: A New Approach to Foreign Aid*. Washington, DC: Center for Global Development.
11. Randomized control trials (RCTs) are helpful but often not appropriate tool for rigorous measurement due to practical considerations, their often prohibitive cost, and ethical concerns about withholding treatment to control groups for the sake of learning.
12. Instiglio (2017) A Practitioners Guide to Results-based Finance: Getting to Impact. https://instiglio.org/wp-content/uploads/2021/02/RBF_PractitionersGuidebook_Instiglio_18Oct2017.pdf Accessed 14 Aug 2022.
13. Author's analysis based on data in: Instiglio, GPBRA (2018) A Guide for Effective Results-based Financing Strategies. https://instiglio.org/wp-content/uploads/2021/02/Guide_for_Effective_RBF_Strategies1.pdf; and accompanying results-based financing database. Accessed 14 Aug 2022.
14. Gardiner S, Gustafsson-Wright E (2015) *Perspectives on Impact Bonds: Putting the 10 Common Claims About Impact Bonds to the Test*. Brookings

Institution. https://www.brookings.edu/blog/education-plus-development/2015/09/02/perspectives-on-impact-bonds-putting-the-10-common-claims-about-impact-bonds-to-the-test/. Accessed 15 Aug 2022.
15. Dalberg analysis.
16. For more information visit: https://www.ubs.com/global/en/ubs-society/philanthropy/optimus-foundation/what-we-do/win-win-situation.html
17. For more information visit: http://www.amazonfund.gov.br/en/home/
18. For more information visit: https://www.pilotauctionfacility.org/content/about-paf

15

When, How, and Why to Use Impact Bonds

As discussed briefly in the previous chapter, an unfortunate simplification often undermines the use and scale up of results-based financing mechanisms. Many still equate all forms of innovative finance or results-based finance with just one flavor of contingency-based financing: the impact bond. However, not only are impact bonds not the only type of results-based finance, but they are also inappropriate in many contexts. I learned this the hard way through personal experience.

A hammer looking for a nail

At a recent working group session organized by the International Red Cross, I had a chance to listen to the successes and challenges of several efforts to design and launch impact bonds for a wide variety of humanitarian and health causes. I was there to share my experiences designing and conducting a feasibility assessment of an innovative financing mechanism to end the plague of cholera in Haiti—a disease unfortunately reestablished in Haiti by United Nations peacekeepers helping with the 2010 earthquake.[1] At the end of the project, we had not only proven the feasibility of using results-based finance but also secured potential implementation partners: Partners for Health, Zanmi Lasante, and PAHO. In addition, we had structured both an impact bond and a time-staged pay-for-success contract and begun to

K. Hornberger, *Scaling Impact*,
https://doi.org/10.1007/978-3-031-22614-4_15

socialize the concept with potential investors and outcome funders. That was when we hit a snag.

Despite the upfront commitment from USAID and ongoing conversations with a large potential working capital provider/investor, we could not secure enough donor commitment from other sources to realistically bring the project to life. The true cost of eliminating cholera could not be broken down. If we were to achieve the outcome of zero cholera cases for a year, sufficient commitment for funding would be necessary. Potential donors had three concerns: (1) Why should they have to pay more than the cost of the programs to see the outcomes realized if we were so certain the programs would work? (2) Why couldn't the UN itself just pay for it, since they had created the problem? (3) Why couldn't the Haitian government pay for it from existing donor funding it was already receiving; that is, why wasn't it a government priority?

Despite the struggle to bring the instrument to life as it was originally designed, we found a silver lining. A leading foundation responded to the need and decided to fund Partners in Health and other implementing partner activities directly through a standard grant instead. In other words, the program was established via traditional philanthropic assistance, not via an innovative structure. This was a successful outcome even though it was not the one we had envisioned. That experience revealed lessons about impact bond viability and measurement as well as about the multiple stakeholders involved and required to make it a success.

While both results-based financing mechanisms that were considered—an impact bond and a time-staged PFS contract—could have been successfully used, we learned that each mechanism is optimal in different sets of circumstances, making the combination challenging. Impact bonds often fit best when supporting interventions that show early promise; put differently, they are best at scaling things that are working well but still involve some level of risk in achieving outcomes. Impact bonds also have the clear advantage of being novel, so investors and funders may see additional value in participating in the transaction to build the field's knowledge. In contrast, a time-staged PFS contract may be more effective for maintaining interventions that are already working well; compared to impact bonds, a time-staged PFS contract's target intervention is likely more certain. Because a time-staged PFS contract is a relatively well-explored mechanism, it is less able to attract funding from actors primarily interested in field building, unlike impact bonds.

In addition, we learned that the ideal outcome metrics for an RBF mechanism vary between measuring intermediate activities and ultimate impact, depending on the intervention being funded. The negotiation of outcome metrics and their associated targets is often the lengthiest and most difficult part of the impact bond structuring process. This is in part because of different perspectives about whether outcome metrics should be closer to activities or should be more related to impact. The field has reached no consensus on this. Prior impact bonds we evaluated used metrics that measure intermediate outputs (e.g., number of households receiving seed payment in the Village Enterprise DIB), which allow investors to verify progress more easily toward an ultimate goal and are typically easier to attribute to implementing partner activities. Others have used metrics that purely measure impact (e.g., educational outcomes in the Educate Girls DIB, discussed below), which gives outcome funders more assurance that they are truly paying for results and allow partners to implement interventions more flexibly. But they may be more difficult to measure and attribute to the interventions themselves.

My conclusion is that a good RBF structure will consider and use both types of metrics: those closer to process and those closer to impact. I also learned that having more than one metric is OK. This helps alleviate concerns among investors and implementing partners with varying preferences as well as concerns that a single impact metric (in our case, the elimination of cholera transmission) often won't work in practice. In this way, an impact bond allows for multiple pathways for earning outcome payments while still recognizing the overarching impact goal.

More importantly, I learned that results-based finance continues to attract donors' attention yet fundraising remains challenging. Every donor we spoke to expressed interest in "doing development differently." From one donor's commitment of $1.2 billion to support innovation in development finance to a leading foundation's creation of an outcomes-based finance fund of funds, funders are investing substantial time and effort in exploring innovative finance. Despite this interest, most donors and foundations are still highly cautious about funding impact bonds—including those who have funded them before—due to their time-to-launch, their cost, and their complexity.

For example, the Children's Investment Fund Foundation (CIFF) questioned whether the high transaction costs required to set up the mechanism were worth its purported benefits, mentioning that impact bonds "feel like a hammer in search of a nail." Because of this, donors' interest in innovative finance often does not translate into actual funding commitments. As we saw in the previous chapter, the value of actual disbursements for RBF remains less than 7 percent of all development assistance. This means that securing outcome funding requires sustained effort and often high-level political engagement from an influential champion. Sometimes the cost is too high to justify bringing the impact bond to life.

Similarly, investors, especially commercial ones, often still have doubts about impact bonds, but creative risk mitigation could help attract more interest. While it is often assumed that "investors will come" to an impact bond once outcome funding is secured, my experience suggests this may not be the case, particularly for more commercial investors. Many feel that impact bonds are risky and unlikely to meet their commercial return requirements. They also need to align with the investors' social impact strategies, which are often varied and may diverge from the specific mandate of the individual impact bond. Moreover, some investors that invested in impact bonds in the past are shifting away from the model due to mixed results. That said, many investors remain interested in exploring impact bonds; their primary concern is how to mitigate the risk. This creates opportunities for creative risk mitigation techniques (e.g., through insurance, scaled outcome targets, or guarantees). Yet impact bond structures should be clear-eyed, recognizing trade-offs of additional risk mitigation given that too much risk protection could undermine the purpose of the impact bond and add unnecessary complexity to the structure.

Given these very real and hard lessons, how and when should one use an impact bond in practice? Let's discuss each issue in turn.

How do impact bonds work?

As mentioned in the last chapter, impact bonds are results-based contracts in which one or more private investors provide working capital for social programs implemented by service providers (e.g., NGOs) and one or more outcome funders (e.g., public sector agencies, donors, etc.) repay investors both their principal and a return if, and only if, the social programs succeed in delivering results (See Fig. 15.1 for details).

Impact bonds sit in middle ground between early-stage interventions (requiring "angel" philanthropic funding) and well-proven models ready for scale (suited to large donor funding). Impact bonds can be a valuable tool to facilitate intensive fine-tuning for those early-stage interventions that demonstrate strong initial results. The collaboration among the implementing organization, donors, investors, and the performance manager in the context of a single-minded focus on outcomes can be a powerful generator for new ideas during this middle stage. Impact bonds also present an alternate

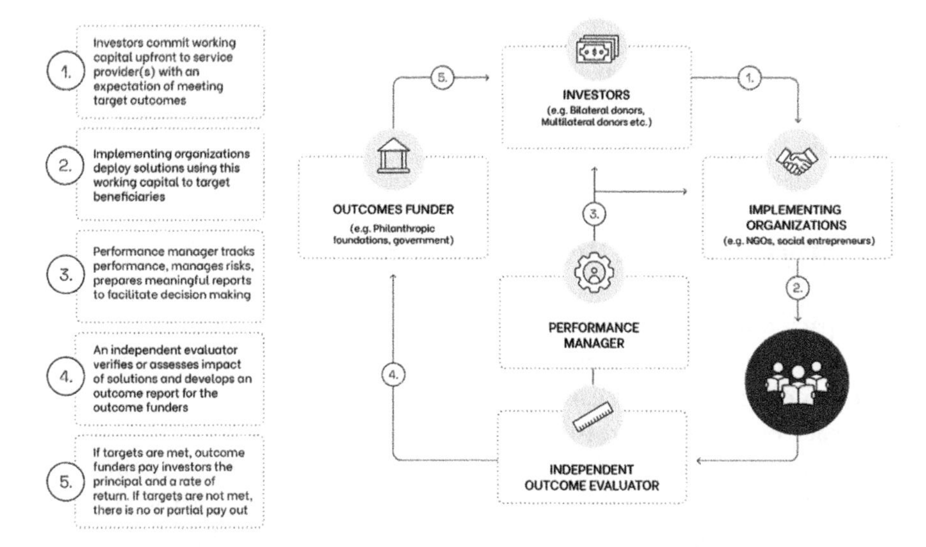

Fig. 15.1 Illustration of actors and relationships in impact bond structure

IMPACT BONDS ARE MORE APPLICABLE WHEN:	IMPACT BONDS ARE LESS APPLICABLE WHEN:
Projects yield clear and measurable social outcomes that can be attributed to the project	Outcomes are almost certain once activities are funded
There is a reasonable timeline to achieve outcomes	There are strong regulatory barriers hindering intervention flexibility and ability to innovate
Relevant outcome data is available and data flow is reliable and timely (including baseline data)	Projects do not yield clear and measurable outcomes
External risk capital is required, especially for projects aimed at extremely marginalized populations	It is difficult to attribute impact to the intervention as opposed to external factors
Service providers are able to meet the challenges and rigors of the impact bond models	There is a lack of commitment from the involved partners
Interventions are innovative but previously demonstrated	The size of tthe problem is not big enough to justify DIB design cost, M&E and private investor pay-out

Fig. 15.2 When to use an impact bond

funding pathway for involving investors with greater flexibility in where and how they choose to invest their money after accounting for the risk-adjusted return.

For those considering impact bonds, the checklist in Fig. 15.2 offers twelve screening criteria that will either help refine the development impact bonds program design or point to other, more appropriate, RBF mechanisms.

Impact bonds will not fit every development problem, but since they can lead to better results produced more efficiently than many alternative approaches, they are worth considering. When used in the right context, the approach can create space for more innovation, local problem-solving, and increased transparency for target beneficiaries and taxpayers in donor countries.

Five myths about impact bonds

As impact bonds gain traction as a means of financing, many misconceptions and myths surrounding them must be dealt with. Here are some of the corresponding truths that can dispel those myths.

Impact bonds are increasingly gaining prominence as financing mechanisms for a wide range of development activities. According to the Brookings Institution, as of July 2022, 235 impact bonds were in place across 38 countries, totaling more than US$463 million in upfront funding.[2] While the popularity of and interest in impact bonds has grown, conversations around them show lack of understanding of how they work and when they can be most useful. The following discussion addresses the truths behind five commonly expressed myths. By remaining aware that others may harbor one or more of these misconceptions, funders, investors, and implementers will be able to use impact bonds more successfully.

Impact bonds are not bonds

The use of the word "bond" misleads development practitioners into thinking that impact bonds resemble instruments used in financial markets to raise money as a loan agreement. In fact, impact bonds are neither bonds nor financial instruments at all. Rather, they more closely resemble multiparty contracts for sharing risk and paying for social outcomes.

Impact bonds can be structured either as social impact bonds (SIBs), with outcome funding provided by the government, or as development impact bonds (DIBs), with funding from the development community (a foundation or philanthropist).

Impact bonds do not crowd in new money

Impact bonds can be crucial tools for driving innovation and risk sharing, but they rarely bring truly new money to the table. One of the key ways in which impact bonds create value is by unlocking additional funds from private equity investors, but this is only true in a minority of cases. Typically, an impact bond requires the outcome funder—a government or philanthropic entity that funds development work in various ways—to provide a grant covering the costs of achieving outcomes, interest for the investor, and fees to other third-party partners that structure, measure, and manage the outcomes achieved.

Impact bonds may not bring new money to address social challenges, but they can be useful in driving innovation and creating value for both risk-adverse outcome funders (those who want guaranteed outcomes) and for investors willing to take on the risk for an appropriate reward. Therefore, it is crucial to ensure a balanced level of risk between funders and investors, such that both parties find the terms of the bond and the risk level acceptable.

Impact bonds are not the only type of results-based finance

As I now have mentioned multiple times, impact bonds are one of many results-based financing (RBF) tools available, and each of them caters to a different set of parameters and priorities. See the previous chapter for more details.

Impact bonds are not suited to all organizations

Many nonprofits with good reputations and track records of success are willing to take on the role of service provider as part of an impact bond. While some for-profit service providers are risk takers, either fully or partially, the vast majority can't ever lose financially because it is always either the investor or the outcome funders who bear the financial risk. Thus, service providers may need additional incentives to drive them to achieve outcomes such as by linking funding to success or risking a loss in reputation. For example, while designing an impact bond, we considered working with public hospitals in India as service providers. However, these hospitals did not have the right incentives to perform better with an impact bond structure than without one, undermining the likelihood of achieving results. Most nonprofits have an incentive to deliver outcomes even if they are not losing financially because their reputation is at stake, and loss of reputation can hamper their fundraising abilities in the future. Public hospitals, however, are not as sensitive to such reputational risks because they will continue to get funding from the government, regardless of their performance on the impact bond. Moreover, the demand for their services from low-income populations often exceeds supply in any case.

Service providers must also have a data-driven decision-making culture and the ability to pivot and innovate based on lessons learned during implementation. Under the impact bond model—where payment is contingent on innovation, learning, and results—an inability to collect data and adapt programs when there are signs of failure can be costly. Thus, it is critical that service providers think carefully about their incentives before entering into an impact bond structure to fund their programs.

Impact bonds can work even when external risks are large and hard to manage

The core value of the impact bond mechanism is to transfer the risks of a development program to someone who can better absorb and manage them. Internal risks, such as the ability to hire on-the-ground staff to deliver the program, are more straightforward to price in and manage. Certain external risks can also be managed or outsourced further. For example, the risks of external stakeholder disruptions (such as from a teachers' union) may be anticipated based on past experience, while weather risk for an agricultural impact bond may be mitigated by purchasing weather-based crop insurance.

However, some external risks are simply not predictable or preventable, such as the COVID-19 pandemic or changes in regulations outlawing a certain funding model, leading to suspension or termination of the agreement under a force majeure provision. If ongoing costs arise despite suspension or termination of contract, the investor and outcome funders would need to renegotiate who will take responsibility, as nonprofit service providers may be unable to do so. In the pandemic scenario, for example, laying off economically vulnerable on-the-ground staff would not be advisable, and nonprofits might not have extra reserves to cover their salaries. A well-designed bond should be able to price in higher returns for taking on higher risks and allow an exit in scenarios where the risk cannot be priced in or managed.

Being more clearheaded about the misconceptions discussed above and not assuming impact bonds are a solution to any type of development, social impact, or environmental problem (a hammer in search of a nail) can allow successful use of this tool. In fact, there are now an increasing number of success cases. One of them is the Educate Girls Development Impact Bond in India, a project with which Dalberg, along with several partners, was fortunate enough to be involved. Since we started with some lessons from a failure in using impact bonds in practice, let's look at lessons from a success story.

An impact bond success story

One of the best-known examples of the successful use of impact bonds comes from the world's first-ever development impact bond (DIB), which focused on girls' education outcomes in Rajasthan, India. The DIB was structured in 2015 and provided US$270 thousand in funding to Educate Girls, an Indian nonprofit working to increase the number of girls enrolled and learning in school. The outcomes the program sought to achieve were a 79 percent increase in enrollment among out-of-school girls across 166 school catchments and improved learning gains on the Annual Status of Education Report (ASER) testing tool for children in grades 3 to 5. Initial investment capital was provided by UBS Optimus, a Swiss foundation. The London-based Children's Investment Fund Foundation would reimburse UBS Optimus's initial investment and pay returns if Educate Girls met or exceeded a set of predetermined goals. The consultancy Instiglio managed the design, IDinsight performed the results evaluation, and Dalberg performed the process evaluation.[3]

In July 2018, IDinsight announced the results of its evaluation of the program. Educate Girls had achieved above its target outcomes: 92 percent rather than 79 percent of all eligible out-of-school girls had been enrolled in the 166 school catchments and a 60 percent above target performance on the ASER test had been achieved. As a result, UBS Optimus recouped its initial funding plus a return of 15 percent on top. More importantly, thanks to the focus on results rather than process, Educate Girls had shifted its model during implementation to focus on group-focused learning rather than classroom-focused learning, with each group based on the children's competency levels rather than their ages.[4] Additionally, Educate Girls provided after-school support, with community volunteers proactively meeting with students and parents outside the classroom to address specific concerns, and focused on outreach to improve enrollment among students in hard-to-reach districts.[5]

These great results were not easy to achieve, and one of the hardest challenges was implementing the evaluation itself. This effort led to many learnings relevant for design and rollout of future impact bonds.

One of the most important learnings was that the impact bond structure is only as effective as the incentives and processes followed to implement and evaluate its performance.[6] Setting ambitious targets based on existing baseline data can help spur innovation. It is also important to provide capacity-building support to the implementation partner, in this case Educate Girls, so they value and understand the goals of the impact bond and how its performance will be assessed. This also helps to ensure that detailed outcome evaluation data is collected and reported in a timely manner. Finally, commitment and trust from investors and outcome funders are critical to both allow for adaptations in the delivery model and to have patience to wait and see if results are achieved after the final evaluation is complete.

The Educate Girls impact bond also offered lessons for one of the most widely held criticisms of the efficacy of the tool—that the cost of structuring them is not worth the outcomes they achieve. In this case, considering the size of the financing need and the costs involved to have multiple parties structure and evaluate its performance, the criticism was warranted. However, it was also a pilot case and offered learnings and successes that have subsequently been replicated for similar impact bonds in India and elsewhere, calling into doubt whether we can fully appreciate the true benefits and costs of individual impact bond transactions.

Nevertheless, to truly scale impact bonds and RBFs on a wide scale, transaction costs must come down, especially for design and structuring, term sheets, and evaluation, but also to increase the quantity of and access to pools of concessional financing willing to serve as outcome funders for replication and scaling of successful models. This will make impact bonds more relevant to a wider group of investors and donors. Some of these actions are already taking place, but more momentum is needed.

Summary of key messages from this chapter

- One type of results-based finance, impact bond, is increasingly being used to reorient donor funding toward results. Nevertheless, these are not optimal for all contexts, and donors and investors continue to show some hesitation in funding them due to poor design of risk transfer, high transaction costs, and lack of familiarity with the tool.
- Impact bonds are, in fact, not bonds but rather results-based contracts. In an impact bond, an investor provides upfront risk capital to an implementor to run a program with predetermined social outcomes. If the outcomes are achieved, an outcome funder pays back the risk capital, plus interest, to the investor.
- Impact bonds are most appropriately used for interventions that are still not fully proven but show promise of achieving desired outcomes where those outcomes are transparent and measurable.
- Impact bonds can work even when external risks are large and hard to manage, as was the case with the Educate Girls Impact Bond in India,

where changes to the learning landscape led to shifts in the program delivery model, which allowed the program to achieve its desired outcomes.

- Impact bonds, and for that matter RBFs more generally, will only be scaled if transaction costs to structure and implement them can be brought down and if the pools of concessional funding to invest in them continue to grow.

Notes

1. United Nations (2016) Press Release: Secretary-General Apologizes for United Nations Role in Haiti Cholera Epidemic, Urges International Funding of New Response to Disease. https://press.un.org/en/2016/sgsm18323.doc.htm. Accessed 15 Aug 2022.
2. Author's analysis based on data in: Brookings Institution (2022) Global Impact Bonds Database Snapshot. https://www.brookings.edu/research/social-and-development-impact-bonds-by-the-numbers/. Accessed 14 Aug 2022.
3. Datla A (2019) Paying to Improve Girls' Education: India's First Development Impact Bond. Harvard Kennedy School Case Study. https://case.hks.harvard.edu/paying-to-improve-girls-education-indias-first-development-impact-bond/. Accessed 15 Aug 2022.
4. IDinsight (2018) Educate Girls Development Impact Bond, Final Evaluation Report. https://golab.bsg.ox.ac.uk/knowledge-bank/resources/educate-girls-final-report/. Accessed 15 Aug 2022.
5. Global Steering Group for Impact Investment. Case Study: Educate Girls Development Impact Bond. https://gsgii.org/case_studies/educate-girls-development-impact-bond/. Accessed 15 Aug 2022.
6. Dalberg (2018) Lessons from the Educate Girls Development Impact Bond. https://dalberg.com/our-ideas/lessons-educate-girls-development-impact-bond/. Accessed 15 Aug 2022.

16

"Link Financial Incentives to Impact"—An Interview with Bjoern Struewer

This chapter summarizes my interview with Bjoern Struewer, Founder of Roots of Impact a pioneer in the development and adoption of the impact-linked finance.

Question 1: How did you first start to get involved in results-based finance?

Bjoern: I was looking for a structural solution to a structural problem. As I entered the impact economy and worked with hundreds of entrepreneurs, I wondered how to address the disconnect between the needs of high-impact enterprises and the realities of capital markets. If these organizations really want to scale significantly and manage to tap into commercial sources of capital, they run the risk of leaving behind the people who need their solutions the most. I felt an urgent need for a simple solution to address the root causes of this pattern. Inspired by impact bonds, I knew that, first, this structure designed for nonprofits was not appropriate for businesses and, second, that a solution for impact enterprises needed to be simpler and more straightforward. I was looking for a way to enable the best companies in the world—in terms of positive impact—to raise the right kind of capital, scale their effective solutions, and further optimize their impact. We need to be clear that any kind of catalytic, patient, concessionary capital involves a subsidy, even if no one likes to use that term. Hence my simple idea: what could be more obvious than to tie any subsidy directly to the achievement of positive impact and to provide it directly to the enterprise creating it? This

K. Hornberger, *Scaling Impact*,
https://doi.org/10.1007/978-3-031-22614-4_16

is how we came to Social Impact Incentives (SIINC)—financial incentives that enable and encourage entrepreneurs to optimize their impact while scaling the enterprise. SIINC is a proven, specific, results-based financing instrument for impact enterprises seeking to raise investment.

Question 2: Why do you think it is important to use results-based approaches to financing?

Bjoern: Creating impact is all about achieving results. And providing financing tied to results is only natural if you want to ensure that those results are really achieved, provided they are measurable and can be clearly attributed to an organization. I don't want to reiterate all the benefits of this approach in terms of improved accountability, effectiveness, and targeting that are evident for nonprofit interventions. For businesses, however, results-based finance has additional benefits. Business leaders and entrepreneurs are accustomed to striving for results—it's in their DNA. They make no excuses if they don't generate enough income and profit to sustain the business. In return, they have some autonomy and flexibility; they can innovate and optimize. It is entirely up to them how they achieve the results. For this reason, it is quite easy to convince an impact entrepreneur that payments or rewards should be linked to the achievement of results. If those results align with the business strategy, you will immediately have a discussion about realistic expectations and a fair price for the expected outcomes. I consider the impact of any enterprise as a performance that can be assessed, managed, and optimized, and I strongly believe that this impact performance can be affected by the way in which the enterprise is financed.

Question 3: What led you to start roots of impact?

Bjoern: It was a desire to fill a significant gap in the market. After leaving the banking industry, I took a sabbatical and spent a lot of time with inspiring entrepreneurs who create positive change with breakthrough innovations. I supported many of them in raising impact investment and even cofounded a social finance intermediary focused entirely on this. During that time, I experienced firsthand the challenges high-impact enterprises face in raising capital and staying true to their missions. As a former banker who is passionate about impact, you would probably have expected me to launch an impact fund. However, there was no need for me to do the same thing that many others were already doing. I was more interested in going where the supply didn't yet meet the demand and where I could be most "additional." I knew that to make a difference and change the practice of impact finance, you had to be on the side of the capital. So, after founding Roots of Impact, my team and I began advising family offices, foundations, and even corporations on "better" impact investing strategies. I wasn't completely satisfied, however, because most impact investors still want to use the same investment paradigms as they do in the mainstream finance world. But my point was to change these paradigms. This led me to recognize the potential of blended and innovative finance and to develop new solutions in this space. Today, we work with public and catalytic funders, as well as with very ambitious impact-first investors. We consider ourselves managers of catalytic capital. As a pioneer in impact-linked finance, we also deeply believe in aligning capital with incentives to drive change for people and the planet.

Question 4: What are SIINC, and how do they work in practice?

Bjoern: SIINC stands for Social Impact Incentives. These incentives are a financing model that rewards impact enterprises with time-limited payments when they achieve positive outcomes. The outcomes, which are independently verified, can take many forms, such as increased income for the poor, greater gender equality, less plastic waste, or better learning outcomes for children. SIINC provide an additional source of revenue and allow these enterprises to improve profitability and raise (further) investment. Actually, the enterprises need to successfully close an investment round in parallel to be eligible. SIINC thus leverages public or philanthropic funds to encourage investment in underserved markets with high potential for positive impact. This feature makes a SIINC a blended finance instrument. From when we pioneered it in Latin America with the Swiss Agency for Development and Cooperation (SDC) in 2016 to today, many examples around the world have demonstrated how SIINC can be used for impact in sectors such as health, off-grid energy, agriculture, employment, financial inclusion, water and sanitation, and tech. We have also learned a lot in recent years about how to optimize the use of the instrument and by designing it in such a way that it becomes as effective as possible. In practice, SIINC payments are made for a limited period of time that starts after the organization has raised investment—usually between two and four years. The expectation is that by then the organization will have reached sufficient scale and/or become financially sustainable and able to sustain and deepen its impact.

Question 5: How do SIINC compare to other types of results-based finance, such as the impact bond?

Bjoern: They appear to be similar and are often confused, but if you take a closer look, there are only two similarities: Both are outcomes-based and both involve investment. Everything else about a SIINC is more similar to carbon finance, with the important difference that SIINC are applicable to all the other SDGs, too. SIINC provide carbon finance for social impact, so to speak. First of all, SIINC is for market-based organizations, not for nonprofits. The payments go directly to the enterprise that creates the value, not to the investor. It's a way to monetize positive externalities and can serve as an additional revenue stream for the enterprise. We use it as a blended finance instrument that leverages repayable investment. The leverage depends on the context, but typically it is between 1:3 and 1:5. In an impact bond, however, the donor pays the full cost of the intervention plus the investor's return in the event of success. Considering that SIINC only pays a business for marginal, incremental impact, it is very cost-effective for the outcome payer. We also design the transactions so that the company is still able to deliver the impact long after the SIINC transaction has ended. I don't mean to say that between SIINC and impact bonds one is better than the other: they are two very different tools for very different purposes. Impact bonds aim to test innovation by shifting risk to investors, while SIINC is used to scale what works and to unleash the full impact potential of market-based solutions.

Question 6: What are some examples of successful uses of SIINC? What results were achieved?

Bjoern: While there are many ongoing SIINC transactions and even more in the pipeline, only a few of them have as yet been completed and fully evaluated. A good, scientifically evaluated example is the SIINC transaction with Clínicas del Azúcar. CdA is the largest private provider of specialized diabetes care in Mexico and serves primarily lower middle-income groups at 40 percent of the price charged by alternative providers. The company's biggest challenge back in 2016 was finding the right investors who would support its ambitious scaling plan while allowing it to move to even lower income segments and ensuring high-quality treatments for all of its patients. With SIINC, we incentivized CdA to increase the proportion of low-income patients. At the same time, we incentivized high-quality treatment so as to achieve measurable health improvements across all patient segments. The total payment amount was capped and spread over a predetermined 2.5-year period. An independent evaluation of the SIINC with Clínicas del Azúcar found that the company had achieved both: its low-income client outreach and its quality targets. Results on low-income patient penetration rates show a significant increase (from an initial 32 percent to 37 percent of total patients). Finally, the results regarding quality of services are significant for both low-income and existing patients. Another great example is

Root Capital, a pioneering impact lender to agricultural businesses in Latin America and Africa. The SIINC enabled Root Capital to make 39 high-impact loans to 32 small, early-stage agribusinesses that would have been unlikely to receive loans on similar terms from existing financial institutions. In this case, SIINC metrics were all about additionality and included a bonus for financing gender-inclusive agribusinesses. Extensive case studies are available on both examples.

Question 7: What are some of the failures or lessons learned you have had along the way as you grow the SIINC product?

Bjoern: When you innovate and test completely new approaches, you have to accept failure, if not provoke it. It is simply important to fail forward and learn lessons from it. While I don't recall any significant failures, certainly many lessons along the way helped us refine our approach to structuring SIINC, selecting suitable companies, and managing respective funds. For example, we learned that SIINC works much better when applied to an entire organization, spurring the enterprise to greater impact. Project-type interventions that are limited in scope and time are not optimal. SIINC was never intended to be used in project finance. We are very happy to work with entrepreneurs who are committed long term and have their skin in the game. We've also learned that SIINC is often confused with traditional grants—and, of course, with impact bonds. For this reason, we need to clearly distinguish SIINC from anything related to traditional grantmaking. SIINC never pays for any activity of the organization or covers any costs for specific activities. Instead, it is a bilateral agreement that rewards impact performance, regardless of how it comes about. The enterprise must agree on a set of relevant metrics that will be used to trigger payments. These metrics are designed to capture its most relevant impact areas. At the same time, they should also be useful for business operations. So when metrics are rewarded, the incentive is to focus on improving them. Already during our first program with SIINC, we realized that SIINC is not just for small, early-stage companies struggling to raise appropriate investment. It is also a great fit for commercially viable companies with impact-conscious leaders who are looking for ways to increase their social impact. Such a scenario is especially valid when a model has achieved, or has the potential to achieve, strong financial performance and scale. These companies often attract a lot of interest from investors. However, they may come under pressure to focus on more lucrative markets with lower-hanging fruits that promise short-term profits—a danger that SIINC can address.

The company is encouraged to focus its business model on maintaining and deepening its impact sustainably.

Question 8: How did you move from SIINC to impact-linked finance?

Bjoern: As we experienced the uptake of SIINC and anticipated its future potential, we quickly realized that there is no reason why the SIINC principle should not be integrated directly into any repayable investment instrument—two in one, so to speak. Indeed, rewards for positive outcomes can be built into all financial instruments, from equity to debt to guarantees. Think of linking the interest rate of a loan to a predetermined impact performance—this effectively lowers the cost of financing and creates strong incentives for the borrower to perform on impact. Simply put, this is an effective way to "bake" impact into the core of financing. We call this practice impact-linked finance and sometimes refer to it as "better terms for better impact." We developed specific design principles for it so that it can be applied by anyone who wants to unlock the full impact potential of a business. The more social or environmental value a company creates, the lower its cost of capital. And, more importantly, it can further optimize its impact. As a result, resources flow to what matters most to society. You might say that this is too bold a vision. But everything is already in place to make it a reality: Many development finance actors and catalytic capital providers around the world are eager to create deep impact and social value with their funds; a growing number of impact investors and venture capitalists are interested in accessing high-quality impact deals; and many great, bold, and inspiring entrepreneurs are ready to scale their proven, high-impact solutions. Given the daunting challenges we face on this planet, it's not just Roots of Impact and a few like-minded players that should be implementing this practice. Every financial intermediary, every consulting firm, and every professional working on impact should be doing it. We need a much bigger push, and it starts with capital. This is why those at the forefront of impact, i.e., the development funders, public agencies, and catalytic funders, are uniquely positioned to push the boundaries.

Question 9: What are the biggest challenges to scaling impact-linked finance?

Bjoern: Impact-linked finance requires enterprises to reliably measure their social and environmental impact. If they can't do that yet, we need to help them. It is critical for a company to have a good understanding of the value it creates for its customers and stakeholders. Done right, this creates multiple, huge dividends. Of course, it would be much easier if we could build on existing systems and standards. In this regard, we applaud the great work of others in establishing standards for measuring and managing impact. Technology is also very helpful. In terms of the bigger picture, an entire study looks at how to scale Impact-Linked Finance: In 2018, we partnered with BCG to analyze what it would take to accelerate this practice. We found that the main barriers are lack of experience, knowledge, capacity, and data. This is not surprising given the short time the practice has been in use. In addition, there may be regulatory constraints with respect to the legal form and setup chosen by a funder. And, last but not least, this new practice has to be attractive for all stakeholders involved: outcome funders, investors, and enterprises. In response to this, and based on our practical experience, we have concluded that we need to make it much easier for funders and investors to get started. That's why we created a one-stop solution, the Impact-Linked Finance Fund. The Fund, established as a Dutch nonprofit foundation, acts as a capital provider and knowledge hub for the practice of impact-linked finance. It pools funding from various sources and manages sector-specific, regional or thematic impact-linked funds from A to Z.

Question 10: What is next for Roots of Impact, and what are you most excited about working on moving forward?

Bjoern: I think the simple idea of "better terms for better impact" is not just another results-based finance approach but a paradigm shift. It has the potential to change the way we finance impact. It could even have positive spillover effects on the way we use finance in general. In any case, I'm pleased to see that many practitioners have begun to incorporate impact into finance in meaningful ways. If you just look at the last few years since SIINC was introduced, there are many signs that the idea of rewarding impact is gaining traction. I want to see this practice grow and take off. And we will do our share by actively building the field with others. One specific step in this direction is that we have launched the Open Platform for Impact-Linked Finance, a knowledge and resource hub to build a global community of practice. We want to grow the group of like-minded pioneers and take impact-linked finance to the next level. In addition, I am excited about some bold ideas that my team and I will be working on. For example, we believe that every impact fund must have an impact incentive facility attached to it so that it can provide better terms for better impact in every single investment. We also believe that impact incentives must be an integral part of any future technical assistance to impact enterprises: In our view, funding for activities should be complemented, if not replaced, by funding for results. It's time to align capital with incentives for positive impact, and I can't think of anything better to work on with all the passion I have.

Part VI

Provide Capacity Building, Not Just Capital

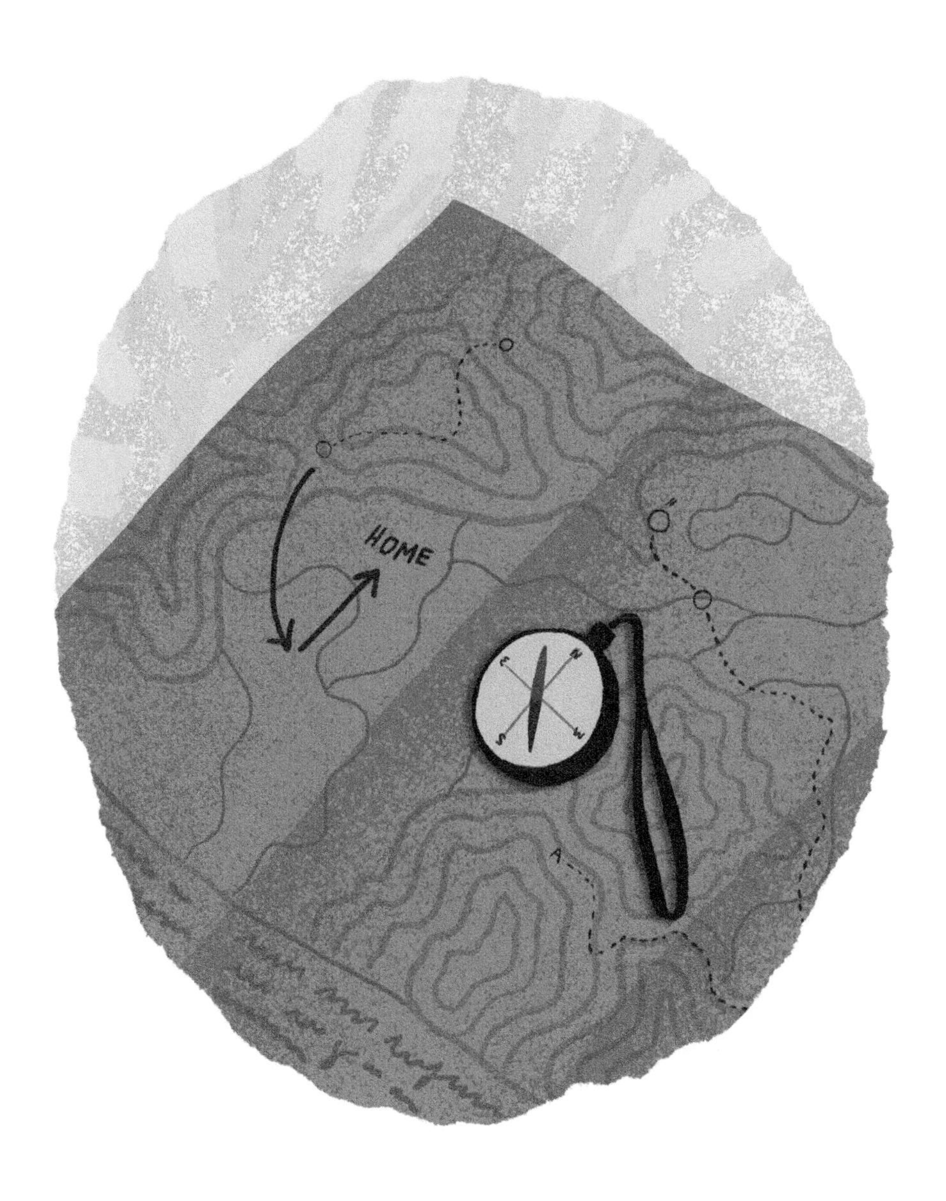

17

Enhancing Impact with Capacity-Building Services

"If you give a man a fish, you feed him for a day. If you teach a man to fish, you feed him for a lifetime." One of my favorite proverbs, this adage is also true for the world of finance and investment. Providing capital alone is often a necessary but insufficient solution.

Throughout various sections of this book, we have seen the centrality of nonfinancial factors in helping to scale capital for impact. In section one, we looked at bundling technical assistance with microcredit or other types of financial services for microentrepreneurs or microenterprises, and in section four we considered how using technical assistance facilities as a type of blended finance can succeed in achieving the SDGs, when capital alone is not enough.

I saw this clearly on my very first paid job after finishing the Peace Corps. As part of TechnoServe Inc. (TNS), I was working with coffee farmers in Tanzania and saw the power of technical assistance firsthand. Working with a collective of farmers who had left their coffee plants to fallow over frustration with the fixed below-market prices at which the government offered to buy their coffee, TNS convinced them that another way was possible.

Before we could secure the capital we knew the farmers needed, we offered collective pre-investment technical assistance on governance and dealing with the operating environment. To do this, we helped a group of pioneering farmers start a small, innovative specialty coffee company. Roughly 300 small-holder farmer families, most living in poverty in the highlands of Tanzania, served as its co-owners. Our goal was to help these farmers improve the

K. Hornberger, *Scaling Impact*,
https://doi.org/10.1007/978-3-031-22614-4_17

quality of their coffee by adopting high-quality, certified-sustainable farming practices; organize themselves as a for-profit company instead of a cooperative selling at fixed prices to the government; and expand their market presence to reach international markets by building connections with buyers willing to pay for quality. At the same time, we worked to secure a loan from a social lender—Root Capital—to provide the group with working capital to upgrade their production equipment and practices. We also worked to catalyze regulatory changes at the Tanzanian Coffee Board and tax reforms for small businesses, allowing the company to grow.

The resulting coffee company was a huge success, but it still needed technical assistance post-investment. After establishing the for-profit entity and securing new buyers, as well as a loan to improve production capacity, the farmers still needed agronomic technical assistance to make it all work. TNS happily supplied that support in the form of regular visits by agronomists and market access advisors to each farm to ensure that good farming practices were being adopted. Thanks to this support post-investment, the farmers inside the company improved their coffee quality from Class 9 to Class 5. This, combined with a better marketing system, allowed these farmers to get an average 70 percent price premium in international markets. We helped the farmers secure direct contracts with well-known coffee companies like Peet's Coffee & Tea and Starbucks, on which they received 150 percent price premiums. After just two years, the company had grown to more than 15,000 smallholder farmers and had roughly $US5 million in annual sales.

Yet despite this clear case of technical assistance enhancing the business and impact of the loan by a social lender, use of technical assistance—also known as capacity-building support or business development services[1]—is limited. Just 44 percent of blended finance funds or facilities include any sort of technical assistance facility.[2] According to MixMarket data, fewer than 40 percent of microfinance institutions bundle their credit and financial services with any sort of nonfinancial education or business development services.[3] Lowest of all, only 25 percent of impact investors have externally funded technical assistance services.[4] See Fig. 17.1.

Too many social lenders still see technical assistance as just an additional cost that doesn't create sufficient value to justify the cost it adds to programming. Others continue to fear that it is a market-distorting subsidy provided in an inevitably untransparent and noncompetitive manner. In my experience neither of these is true on aggregate though can be at times true in specific cases. It is therefore critical that capacity-building support is well-designed and executed.

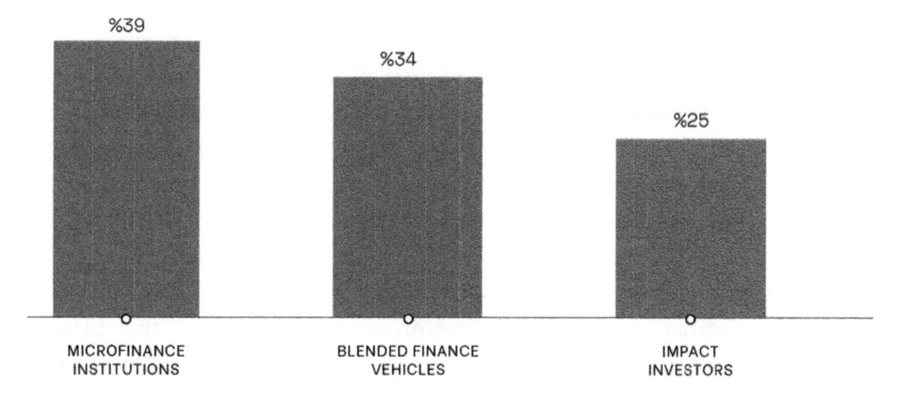

Fig. 17.1 Share of capital providers also providing capacity-building services, 2022 (*Source* Convergence, GIIN, MixMarket)

Making the case for capacity-building support

Considering the important role that micro, small, and medium-sized enterprises (MSMEs) play in low- and middle-income countries, it is essential that they become focal points for all efforts to achieve the SDGs. MSMEs create jobs, drive innovation, and deliver essential products and services to billions of consumers across the globe. In fact, MSMEs contribute an estimated 60 percent of new jobs in emerging markets[5] and have the potential to grow their contribution to GDP from 30 to 50 percent.[6]

While access to finance and additional investment will be essential to realizing the potential of MSMEs, finance and investment alone will not be enough. Capacity-building support provided by accelerators, incubators, technical assistance providers, and donors, along with other entities, plays an important role in helping many MSMEs to reach their potential. Capacity-building services consist of such nonfinancial support as training, mentoring, consulting, networking, and coaching. Effective support services help enterprises generate additional revenue, improve productivity, create jobs, and ultimately improve the livelihoods and well-being of the populations in emerging markets.[7],[8],[9] Years of research have identified effective approaches to capacity-building services across a variety of enterprise segments and types of services.[10]

For example, while initially disputed, entrepreneurship training has modest but significant effects on improving business practices and business outcomes for microenterprises.[11] Additionally, consulting and sector-specific professional advisory services appear to work, leading to improvements for both medium/large-sized enterprises and micro and small enterprises (MSEs). Consulting is expensive, however, and it is less clear how such programs can be scaled, especially for smaller enterprises. Group-based models offer potential.[12] Finally, matching enterprises with well-performing peers also offers promising results, although the impacts depend on the type of peer, and only certain information will diffuse this way, which could limit the contexts in which this approach can be effective.[13]

We have also now clearly seen the disproportionate benefits of offering support services to enterprises in traditionally marginalized populations. Training programs integrating gender-specific content, for example, tend to have positive impact on enterprise performance. Both traditional business training and socioemotional training programs have had positive impact on the performance of women-owned businesses. As in MSMEs generally, particular peer or social networks are likely to play a central role in female-led enterprise performance overall.[14]

Not all capacity-building support is equal

Unfortunately, evidence also shows that not all enterprise performance support works. For example, evidence on the effect of incubators and accelerators in low- and middle-income countries is still limited, although the evidence that does exist suggests that the benefits can be worth the costs in some limited situations. I have been approached countless times with the suggestion that accelerators or incubators be used to solve investors' pipeline problems, but rarely has this proven to be the correct solution. Often, these programs generate false expectations for entrepreneurs and fail to create the type of growth or fulfill the dreams that many accelerators promise.

My experience and the evidence strongly suggest considerable variation in the performance and effectiveness of enterprise capacity-building services. For example, a landmark study conducted by the Global Accelerator Learning Initiative (GALI) in 2016 demonstrated enormous variation in performance across programs. On average, the highest-performing programs generated nearly $60,000 more in incremental investment growth of participating enterprises than the lowest-performing programs.[15] Further studies conducted by Village Capital,[16] USAID,[17] and World Bank[18] all point to the same conclusion: high variability in enterprise outcomes even for programs delivered by the same provider and using the same methodology.

The factors driving this variation are not always clear, but again evidence is emerging that some types of enterprise support and practices work better than others. Results also vary considerably according to the context and the type of capital provider that is offering the capacity-building services. Before digging deeper into the evidence on what works, the rest of this chapter looks at different types of enterprise-focused capacity-building services and delivery models. Chapter 18 will explore in more detail the most consistently effective practices.

Types of capacity-building services

Capacity-building services go by various names in different contexts, such as enterprise support, business development, or technical assistance. All of these variations essentially use grant/donor funding to provide direct access to goods or services in support of the impact objectives of a specific investment or portfolio of investments. For practical reasons, most finance providers break enterprise support into two types, depending on when it is provided: pre-investment support or post-investment support. According to Convergence data, roughly one-third of technical assistance is done pre-investment, while two-thirds is typically done post-investment.[19]

Pre-investment enterprise services typically include market studies, pipeline curation, project preparation, feasibility studies, environmental and social risk assessments, and implementation and stakeholder negotiations. Post-investment capacity-building support often focuses much more on the enterprise's specific needs. These are most commonly trainings or advisory support on financial management, impact measurement and management, environment, social or governance matters, human resources, product development, market access/marketing, and IT/technology implementation. Post-investment support may also include sector-specific technical expertise, such as agronomic technical assistance or climate emissions measurement advice. According to the GIIN, impact investors address various investee needs through capacity-building support, with the most common uses being talent/human resources, impact measurement and management, specialized technical assistance, accounting/financial management, and environment, social, and governance (ESG) advice.[20] The following list characterizes each form:

- **Human resource/talent development** support includes helping enterprises to improve staff capacity by bringing in and retaining talent as well as by improving human resources policies and training programs.
- **Impact measurement and management (IMM)** support includes helping enterprises to develop impact theories of change and tools for data collection, measurement and reporting practices, and staff capacity building for

implementing and taking value from implementing the new practices and tools.

- **Accounting/financial management** support includes developing robust financial accounting and reporting systems as well as training staff to support better practices in managing cash flow, preparing financial statements, and managing sources and uses of funds.
- **Specialized technical support** is often specific to a sector or business need. This can include getting export or manufacturing quality certifications, improving supply chain management, or improving sustainable farming practices, among many others.
- **Environmental, social, and governance (ESG)** support includes helping enterprises build diverse, effective, and functioning board governance. It also means ensuring strong labor-friendly policies that ensure worker rights as well as helping enterprises understand and mitigate their negative environmental impacts.

In practice, as I often play the role of technical assistance provider, I have worked with commercial banks, impact investment funds, and development finance institutions in all these types of capacity-building support roles. I have discovered that such support is most effective when it centers on enterprise needs. The enterprise will then perceive and take value from the support rather than go through the motions because the capital provider requires it.

For example, on a recent project with an impact investment fund that included support from a donor, we designed a training program for portfolio company leaders to help them survive and thrive through the COVID-19 pandemic. Our initial focus had been on soft skills, leadership, and human-resource management. But consultations with the leadership teams indicated their acute desire to address ESG and learn how to better understand and mitigate the ESG risks within their own companies. In response, we pivoted and designed an ESG-focused training program that proved to be a huge success. At the end of the trainings, the founders all laid out clear plans on how they were going to use the learnings to change how their enterprises operated.

Capacity-building delivery models

Capital providers can offer a broad range of capacity-building services, and the appropriateness and type of services offered can be very context-dependent. In general, at the broadest level, capacity-building services are either provided in-house by the capital provider's existing staff or provided by external advisors or experts.

In-house capacity building is generally done as a natural extension of existing operations and leverages the expertise of the capital provider. It is most frequently provided in the form of management and governance support, by equity investors to founders. For example, loan officers of microfinance institutions may offer borrowers financial literacy training or debt lenders may provide borrowers with risk management advice. Founders of high-growth ventures can be better supported to realize enterprise value through the long-term relationship advice and partnership they receive from equity investors.

Externally provided delivery models are much more diverse. Services generally include various types of business advisers, consultants, mentors, training programs, and accelerator and incubator models. Capital providers may use external advisors or consultants for discrete, time-bound projects requiring specialized expertise, for example, to support a portfolio company in implementing a new IT system or adopting better impact measurement and management practices.

Alternatively, providers may offer access to a pool of senior experts or mentors who can be called upon as needed for industry/sector-specific expertise. One example of this approach is taken by a fund I have advised that invests in medical devices and uses a pool of former medical device executives to advise start-ups on their operations. More generally, some form of executive education, often in partnership with a business school or relevant industry-specific educational institution, may be offered to develop the leadership or technical skills of a supported enterprise. I recently had the opportunity to participate in one such training with the Stanford Byers Center for Biodesign, which was very effective.

The most common form of external capacity building is via partnership with external accelerators, incubators, or entrepreneurship communities of practice. While the evidence is mixed on the efficacy of these models, as previously mentioned, they continue to be widely prevalent. Most often, pre-investment models provide guidance from successful entrepreneurs or business angels who work with other start-up and business growth experts to advise companies on how to launch and quickly grow their businesses.

Digitalization[21] of capacity building

The COVID-19 crisis accelerated a trend toward using digital technologies to deliver capacity-building services to enterprises efficiently, inclusively, and effectively. While the shift toward digital had been gaining ground for years, the pandemic gave it even more momentum and urgency, making it more dramatic than anyone could have imagined. A survey I ran for Dalberg as part of its work with Argidius Foundation revealed that while only 24 percent of the 33 enterprise support organizations we surveyed delivered training digitally pre-pandemic, a whopping 62 percent planned to be primarily digital after the pandemic.[22]

Now, post-pandemic, many organizations must determine what role digitalization will play in their approach to delivering capacity-building services to address enterprise needs. Some will choose to reintegrate in-person service delivery as their core business model, while others will maintain a digital focus as a central feature of their strategy. To help organizations figure this out, in our work with Argidius Foundation we developed three pathways for possible deployment—*Enhance, Shift,* and *Transform*—with each applied according to the organizations' desired strategic weight and level of innovation with digital tools (see Fig. 17.2).

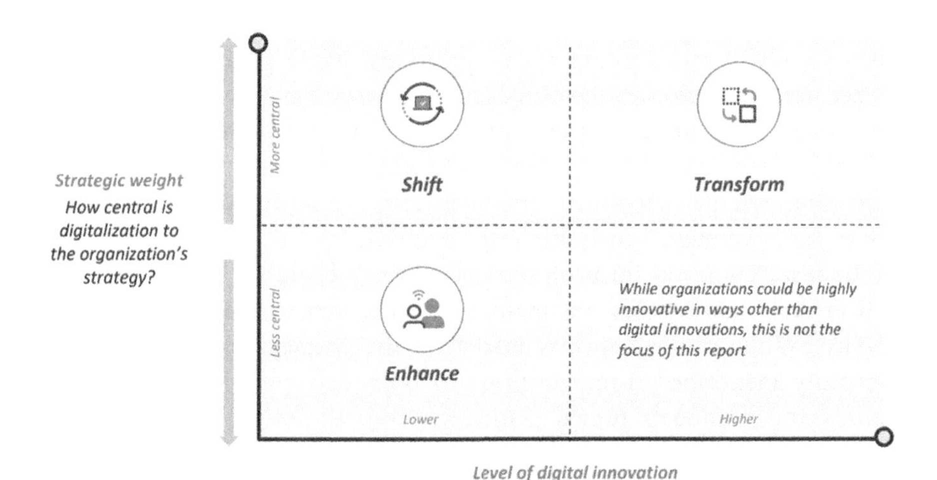

Fig. 17.2 Digitalization archetypes based on two distinguishing characteristics (*Source* Argidius [2020], Digital Delivery: A digitalization guidebook for enterprise support organizations)

Enhance organizations deliver client value through a core offline service model supplemented by digital tools and practices adopted for reasons such as improving service delivery efficiency, but the digital tools are not central to the organizational strategy and business model. Balloon Ventures offers an example. It provides technical assistance to high-potential small businesses in Kenya and Uganda and plans to primarily deliver in-person trainings supplemented by digital tools such as financial management tools and operational checklists. It takes this route because its clients are generally not technologically advanced, and some have limited access to ICT infrastructure, making continued offline support important. At the same time, supplemental digital tools increase the efficiency of their interactions with businesses, enabling them to better identify key areas for improvement as well as dedicate more time and resources to their core in-person capacity-building support.[23]

Shift organizations have a strategic focus on using online digital channels to deliver services. This may be complemented with targeted offline service components, though digital delivery is emphasized. The DO School provides an example. This global organization supports social entrepreneurs, and after the pandemic they plan to offer primarily online workshops with supplementary offline experiences. The DO School takes this approach because they find that online content offers great advantages of scale, inclusivity, and access for their global client base. They will complement this with targeted offline experiences where serendipitous relationship-building cannot be easily replicated online.

Transform organizations have a digital-centric strategy and highly innovative business models, approaches, and tools for digital delivery. Their extensive in-house capabilities and resources enable them to create their own technology platforms or products. An excellent example of a transform model is African Management Institute (AMI), which enables ambitious businesses across Africa to thrive by offering them practical tools and training. AMI does this by offering its own digital platform and content delivery application through which businesses can customize a learning journey with downloadable tools and content and participate in interactive workshops or webinars. AMI's vision was always to build a scalable digital-first model across Africa. Its business model has enabled it to scale through its own services and via licensing arrangements that allow intermediaries to use and customize AMI's web and mobile apps, training content, and methodologies.

Summary of key messages from this chapter

- Capacity-building support, while not present in the majority of transactions, is an important and effective way to ensure that any capital provided, whether through microloans or large private equity investments, achieve its intended impact.
- Despite its considerable benefits, capacity-building support varies widely in effectiveness due to variations in delivery models and types of services as well as enterprise capacity and ability to onboard advice.
- Capacity building includes a wide range of pre- and post-investment support services. Pre-investment, it may include market studies, pipeline building, or technical and feasibility studies. Post-investment, it typically involves some form of operational support to improve enterprise performance, such as through better financial management, risk management, human-resource capacity building, marketing, or better ESG and IMM practices.
- Delivery models include the in-house model, which leverages capital providers' existing capacities and knowledge to reinforce transaction goals, and the externally provided support model, which uses outside expertise/knowledge to enhance performance along one or a multiple of dimensions necessary to realize enterprise potential.
- Organizations are shifting to increasingly digital delivery models, a trend likely to accelerate. Three pathways capacity-building organizations can pursue to become more digital are *enhance, shift*, or *transform.*

Notes

1. Business development services are nonfinancial services offered to help startups, MSMEs, and SGBs tackle obstacles more effectively, speed up growth, and achieve greater scale. These services include acceleration, incubation, technical assistance, coaching, consulting, and other forms of nonfinancial support.
2. Convergence Database: Historical Deals (2022). Accessed 17 Aug 2022.
3. World Bank Mix Market Database: Social Performance (2019) https://datacatalog.worldbank.org/search/dataset/0038647. Accessed 17 Aug 2022.
4. JP Morgan, GIIN (2015) Eyes on the Horizon: The Impact Investor Survey. https://thegiin.org/research/publication/eyes-on-the-horizon. Accessed 17 Aug 2022.

5. World Bank. Small and Medium Enterprises (SMEs) Finance: Improving SMEs' Access to Finance and Finding Innovative Solutions to Unlock Sources of Capital. https://www.worldbank.org/en/topic/smefinance. Accessed 17 Aug 2022.
6. Ayyagari M, Demirguc-Kunt A, Maksimovic V (2014) Who Creates Jobs in Developing Countries? *Small Business Economics* 43: 75–99. https://doi.org/10.1007/s11187-014-9549-5. Accessed 18 Aug 2022.
7. McKenzie D (2017) Identifying and Spurring High-Growth Entrepreneurship: Experimental Evidence from a Business Plan Competition. *American Economic Review* 107(8): 2278–2307. https://doi.org/10.1257/aer.20151404. Accessed 18 Aug 2022.
8. TechnoServe (2019) TechnoServe Is Rated #1 Nonprofit for Reducing Poverty. https://www.technoserve.org/blog/technoserve-is-rated-1-nonprofit-for-reducing-poverty/. Accessed 18 Aug 2022.
9. Bruhn M, Karlan D, Schoar A (2013) *The Impact of Consulting Services on Small and Medium Enterprises: Evidence from a Randomized Trial in Mexico.* World Bank Group. http://hdl.handle.net/10986/15867. Accessed 18 Aug 2022.
10. McKenzie D, Woodruff C (2020) Training Entrepreneurs. *VoxDevLit*. https://voxdev.org/lits/training-entrepreneurs. Accessed 18 Aug 2022.
11. McKenzie D, Woodruff C (2020) Training Entrepreneurs. *VoxDevLit*. https://voxdev.org/lits/training-entrepreneurs. Accessed 18 Aug 2022.
12. McKenzie D, Woodruff C (2020) Training Entrepreneurs. *VoxDevLit*. https://voxdev.org/lits/training-entrepreneurs. Accessed 18 Aug 2022.
13. Endeavor Insight (2018) Fostering Productive Entrepreneurship Communities: Key Lessons on Generating Jobs, Economic Growth, and Innovation. https://endeavor.org/wp-content/uploads/2021/09/Fostering-Productive-Entrepreneurship-Communities.pdf. Accessed 18 Aug 2022.
14. Women Entrepreneurs Finance Initiative (2022), The Case for Investing in Women Entrepreneurs. https://we-fi.org/wp-content/uploads/2022/06/We-Fi_Case-for-Investment_FINAL_light.pdf. Accessed 1 Sept 2022.
15. Guttentag M, Davidson A, Hume V (2021) *Does Acceleration Work? Five Years of Evidence from the Global Accelerator Learning Initiative*. Global Accelerator Learning Initiative. https://www.galidata.org/publications/does-acceleration-work/. Accessed 18 Aug 2022.
16. Roberts P, Lall S, Baird R et al (2016) *What's Working in Startup Acceleration: Insights from Fifteen Village Capital Programs*. Aspen Network of Development Entrepreneurs. https://www.andeglobal.org/publication/whats-working-in-startup-acceleration-insights-from-fifteen-village-capital-programs/. Accessed 18 Aug 2022.
17. USAID (2018) Accelerating Entrepreneurs: Insights from USAID's Support of Intermediaries. https://www.usaid.gov/sites/default/files/documents/15396/Accelerating-Entreprenuers.pdf. Accessed 18 Aug 2022.

18. McKenzie D (2020) *Small Business Training to Improve Management Practices in Developing Countries: Reassessing the Evidence for 'Training Doesn't Work'*. World Bank Group. http://hdl.handle.net/10986/34506. Accessed 18 Aug 2022.
19. Convergence (2019) Blending with Technical Assistance. https://www.convergence.finance/resource/9ncqGGGACPGY9QonSWGSm/view. Accessed 18 Aug 2022.
20. Pineiro A, Bass R (2016) Beyond Investment: The Power of Capacity-Building Support. GIIN. https://thegiin.org/research/publication/capacity-building. Accessed 18 Aug 2022.
21. I refer to *digitalization* as the strategic use of digital technologies, as opposed to *digitization,* which refers more narrowly to encoding analog information to digital format.
22. Argidius Foundation (2021) *Digital Delivery: A Digitalization Guidebook for Enterprise Support Organizations.* Aspen Network of Development Entrepreneurs. https://www.andeglobal.org/publication/digital-delivery-a-digitalization-guidebook-for-enterprise-support-organizations/. Accessed 18 Aug 2022.
23. Argidius Foundation (2021) *Digital Delivery: A Digitalization Guidebook for Enterprise Support Organizations.* Aspen Network of Development Entrepreneurs. https://www.andeglobal.org/publication/digital-delivery-a-digitalization-guidebook-for-enterprise-support-organizations/. Accessed 18 Aug 2022.

18

Fulfilling the Potential of Capacity-Building Services

As we saw in the last chapter, capacity-building services that accompany capital offerings can take many forms, and the providers offering them do so with many goals and motivations. As also noted, capacity-building services vary widely in performance. Many well-intentioned programs unfortunately do not realize the outcomes needed to support enterprise success.

I saw this firsthand through observation of an acceleration program organized by Agora Partnerships in Central America. Agora would identify and bring together a cohort of 30 aspiring entrepreneurs in the region for a bootcamp, held in Nicaragua or Mexico and lasting several days, that included sessions on refining a business model, developing a more investor-friendly pitch, and practicing their new skills with invited potential investors. Agora followed this up by assigning an external consultant with intensive coaching and advisory experience to support the entrepreneurs for four months following the camp experience to help ensure they succeeded. Despite the well-designed and well-intentioned programming, however, most of the participating enterprises were unable to raise funding and ultimately failed.

This was disappointing for both the entrepreneurs and the numerous individuals and organizations that had tried to support them. What could have been done better? Was funding the right objective? Did Agora Partnerships select the right types of enterprises per cohort? Many questions were left unanswered. To its credit, Agora Partnerships learned from the experience and pivoted its model. Today, Agora uses an approach that not only offers better designed bootcamps and consulting support but also broader support

K. Hornberger, *Scaling Impact*,
https://doi.org/10.1007/978-3-031-22614-4_18

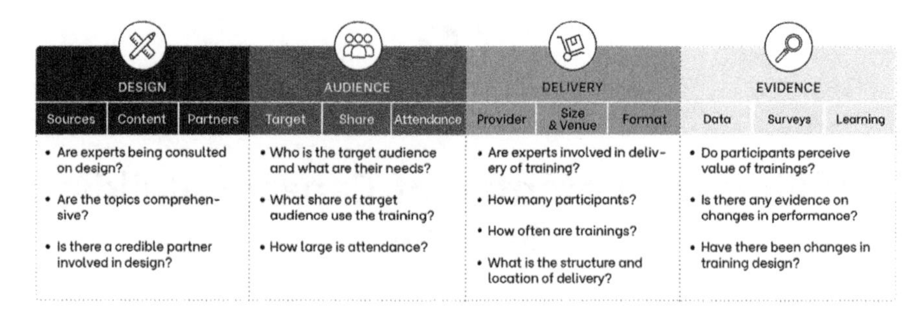

Fig. 18.1 Questions used to evaluate microfinance capacity-building programs

to foster more conducive entrepreneurial ecosystems in Latin America. I hope many other support efforts can be as responsive, as the world does not need more well-intentioned accelerators or incubators that give bad business ideas oxygen while good ideas go underfunded. We need to better design and implement capacity-building services.

In my time working at Global Partnerships, I developed a system to understand what high-performing capacity building might look like for microfinance institutions that offered nonfinancial education to their borrowers. I developed a set of questions that looked at the design, audience, delivery, and evidence of the programs provided (see Fig. 18.1). I found that well-performing programs were designed with experts and partners that not only reached out but also listened to the needs of their intended audience, had regular trainings in formats conducive to learning, and continuously sought to learn from evidence on how the programming was working. Similar to what the academic literature shows,[1] I found that the MFIs providing better trainings had fewer nonperforming loans and grew faster, with higher customer satisfaction. This tool focused specifically on microfinance institutions, so unfortunately, is not applicable to other contexts.

To fulfill the broader potential of capacity-building support services, we need to understand and apply the growing body of evidence on what works to increase the effectiveness of future programming. This is partly the reason that, for nearly a year, I worked with a Dalberg team to support Argidius Foundation in synthesizing its learnings from providing grant funding into the enterprise support ecosystem. Prior to this engagement, Argidius had collected eight years of learning across more than 500 entrepreneur capacity-building programming cohorts that supported more than 12,000 individual enterprises.

We used those learnings, along with a robust literature review and learnings from other peer funding organizations, to gain insights into the most effective types of capacity-building programs. By analyzing patterns in the evidence,

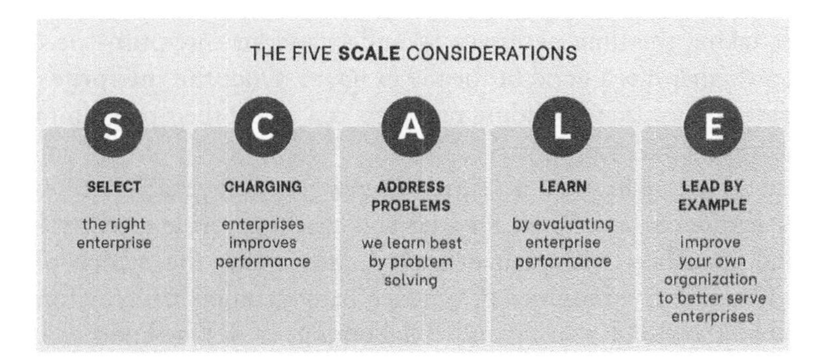

Fig. 18.2 The five SCALE considerations

five key considerations emerged that distinguished cost-effective, impactful capacity-building programs from less effective programs. We codified these, appropriately enough, under the acronym SCALE (see Fig. 18.2).[2]

SCALE serves as a set of helpful considerations rather than a prescriptive framework. It is intended to align enterprise-focused capacity-building providers and funders using a common language and to reveal shared goals so programs can be effectively designed and adjusted around what works to achieve enterprise growth. Although each consideration can be implemented as a standalone component, when deployed together they have the potential to reinforce one another. The remainder of this chapter, drawing on research conducted with Argidius, summarizes each of the five dimensions.

Select the right enterprise to support

Identifying and selecting the right enterprises to participate in a capacity-building program, particularly those focused on pre-investment, is critical. High-performing capacity-building programs are more selective than low-performing programs and tend to target enterprises and entrepreneurs with common characteristics, including enterprises with proven ability to generate revenue, some form of intellectual property, and founders with higher levels of education or management experience.[3]

Capacity-building programs that focus on attracting a small but qualified pool of candidates through referrals or other forms of outreach are likely to have higher-performing programs than those seeking to attract as many candidates as possible.[4]

A strong selection process thus enables capacity-building providers to tailor their program to enterprises' needs more effectively. A good selection process includes taking the time necessary to understand an enterprise's needs and evaluate whether it is a good fit for the program. Once the enterprise's needs are understood, capacity-building providers can adjust their programming to offer the right kind of support.

Selecting the right mix of entrepreneurs can also enable peer-to-peer learning within cohorts. While mentors and coaches provide excellent knowledge and guidance, entrepreneurs often most value the advice of their peers.[5] Creating communities of practice in noncompetitive environments can enable knowledge sharing and collaboration as well as build emotional support among participants.[6]

Careful selection allows capacity-building providers to identify entrepreneurs who are motivated to grow and prepared to make the changes necessary to succeed. Training entrepreneurs to develop skills will have limited effect if the participants are not motivated to expand their businesses.[7] However, some entrepreneurs, particularly from micro and small enterprises (MSEs), haven't yet recognized their capacity to grow.[8] A strong selection process will increase the likelihood that the right kind of motivated enterprises receive the training and support they need to succeed.

Capacity-building organizations can take six actions to help them make better selections of enterprises to participate in their programs[9]:

- Establish clear selection criteria for profiling enterprises and entrepreneurs.[10]
- Leverage referrals from past and current participants and others in the ecosystem.[11]
- Set clear and realistic expectations for enterprises.
- Segment the portfolio and tailor the program approach by segment.
- Use a multistage process to periodically assess partnerships and allocate resources to the enterprises that are most engaged and can derive the most value from the program.
- Integrate feedback loops to inform selection criteria going forward.

One organization I am familiar with that has made careful participant selection central to their programs is Alterna.[12] Alterna supports sustainable, market-driven, context-relevant solutions to development challenges in Central America through locally driven entrepreneurship. In 2019, Alterna identified that selection as a bottleneck to scaling operations. In response, they focused on three aspects: (i) improving segmentation to ensure the right entrepreneurs were matched with the right programs, (ii) building a pipeline team dedicated to attracting and screening entrepreneurs, and (iii) investing in CRM (customer relations management) to establish a back-end system for running the process. As a result, Alterna doubled the number of projects it could run in a year. It also led to customized services into Basic and Advanced, based on the two primary enterprise segments they supported. Having tailored selection criteria for each segment was key to identifying enterprises and matching them to the services they needed.

Charge enterprises to improve performance

A second and somewhat more controversial practice is to charge enterprises to participate in capacity-building programs. Charging is not just about making programs more sustainable but also about finding a price point that encourages engagement and learning but does not unnecessarily exclude participants.[13] Charging for participation enables better outcomes by attracting the right candidates and increasing engagement.[14]

Many capacity-building providers commonly choose not to charge because they believe it goes against their mission, or they assume that enterprises cannot afford the fees.[15] However, a trial in Jamaica explored microenterprises' willingness to pay for training and found that all were willing to contribute at least a nominal fee.[16]

Charging can help capacity-building providers to select the right candidates, and thus increase engagement in the program. On average, only 65 percent of participants attend business training programs when they are offered for free.[17] One program found that participants that paid attended more classes than those receiving the training for free. Charging screens out the enterprises less likely to attend and selects for firms that expect to benefit more.

Capacity-building providers can take the following actions when implementing charging[18]:

- Communicate the value of charging.
- Build relationships based on value.
- Define and test the appropriate level of contribution and payment model.

A key aspect to communicating the value of charging is to clearly explain the benefits to staff and participants of training programs. It is also important to determine a price that correctly balances encouraging active investment against excluding would-be participants. It may take multiple iterations to find the appropriate price point.

After many years of providing free donor-funded capacity-building services to entrepreneurs, TechnoServe now openly promotes self-selection by charging, and it believes the right enterprises see the value of its services and choose to participate and pay for their programming. Charging has allowed TechnoServe to obtain high-quality feedback from its participants, as entrepreneurs expect more when they pay. Charging has also promoted a culture of continuous improvement in the organization, which it finds to be critical to the longevity and success of their programs.

Address problems: focus on group-based problem-solving approaches to learning

The third practice focuses on how capacity-building programs are delivered. Program design that seeks to address the problems enterprise leaders confront, as well as jointly problem-solve with them and understand how to avert those problems, fosters growth and the productivity-enhancing behaviors within the enterprises supported.

Most people learn best by problem-solving.[19] Research has found that traditional business training, focused on teaching a broad range of topics in a short time, has limited effect on enterprise performance (similar to the Agora Partnerships model), whereas approaches that help entrepreneurs diagnose and solve their own problems go further in helping businesses improve outcomes. Consulting services in particular can help enterprises address specific business problems and have been shown to increase productivity, employment, and wages.[20] Alternatively, bringing in experts through outsourcing and insourcing has also been found to improve business outcomes and can be more cost-effective than building those skills through consultation.[21]

Business training can be effective if it provides relevant content. For example, a study conducted in South Africa demonstrated that targeted training in marketing and finance improves profits. More specifically, enterprises with no previous experience benefited from the growth focus offered by training in marketing, while established businesses benefited more from the cost-saving focus of financial training.[22]

Studies have consistently shown that working with teams, peers, and mentors improves business performance more than "in-classroom" activities.[23] In addition, problem-solving in group settings can deliver impact at a lower cost, making it overall more effective. While individual consulting services can positively impact enterprise growth and productivity, cost is a barrier to scaling this approach. Alternatively, group-based consulting, with cohorts of three to eight enterprises, can improve management practices by as much as individual consulting at one-third of the cost.[24] For example, one study showed that a group-based consulting intervention helped increase sales and profits for small groups of 6 to 15 workers in Colombia's auto-parts sector.[25]

Encouraging entrepreneurs who are not direct competitors to regularly interact and discuss business challenges leads to information sharing about suppliers and improved management practices. This approach can encourage enterprise growth and increase the total number of clients and suppliers.[26] Moreover, entrepreneurs grouped with more experienced peers generate faster growth.[27] The opportunity to share solutions and learn from others may be why peer interactions are one of the program components most valued by entrepreneurs.[28]

Entrepreneurs who are part of peer networks exhibit median annual revenue growth higher than the OECD definition for high-growth firms.[29] Fostering such networks can lead to more cohesive and productive entrepreneurial communities. Moreover, networks to support entrepreneurs are most effective when successful entrepreneurs lead the effort and serve as mentors.[30]

Capacity-building providers seeking to build peer and problem-solving-based training services can take five actions to bring these approaches to life[31]:

- Assess enterprise problems accurately by spending time listening to entrepreneurs' problems.
- Build relationships based on trust by promoting personal relationships built on empathy and listening.
- Facilitate learning among peers who face and have overcome similar challenges.
- Make business education relevant, engaging, and easy to apply.
- Build in time to apply the learning.

Two organizations I have engaged with have really taken network-based problem-solving to heart. Enablis has recognized that peer networks are formed by people, not enterprises. Through its work in Senegal, it focuses on getting to know the entrepreneurs, their motivations, their values, and most importantly their stories. Enablis thus builds trust with entrepreneurs, and once the entrepreneurs enter the program, Enablis gives them supported space in which to develop plans based on their actual problems and priorities.[32]

Latitud also uses a peer-to-peer platform to invest in and support tech entrepreneurs to succeed in Latin America. Latitud has built a founder community that doesn't require equity to join but does have a high bar for applicants. Once in the program, participants receive mentoring based on addressing key pain points common to digital products and learning to think in new ways about problems.

Learn by evaluating enterprise performance

One of the biggest mistakes I see capacity-building providers make is failing to build continuous feedback loops and to learn from experience about their training programs' design and implementation. Monitoring, evaluation, learning (MEL) helps providers better understand if enterprises are being supported to grow and how that growth might be achieved more effectively.

Effective performance management offers providers of capacity-building services information about whether supported enterprises are growing, the extent of any growth, and what is and what is not working in their programs.[33] Such information helps providers identify what is driving impact and can help them to understand whether they need to iterate their programs to achieve better outcomes.[34] For example, a randomized controlled trial (RCT) on a social entrepreneurship program in France found that it had no detectable impact on the creation of new ventures, but the findings were used to iterate the program design, and its subsequent model, also evaluated with an RCT, generated a strong impact.[35]

Evidence suggests that capacity-building providers that effectively monitor and use enterprise performance data generate stronger supported enterprise growth over time.[36] Data is critical to decision-making, but not all entrepreneurs know how to extract its value. For larger enterprises, financial training can integrate the use of data to boost profits and efficiency.[37] For smaller enterprises, however, this information may prove too complex to be actionable.[38] For those businesses, training based on rules of thumb is more effective.[39] Data can also help entrepreneurs change course. High-performing entrepreneurs recognize that pivoting can drive growth.[40] However, knowing when and in what direction to pivot requires access to quality data.

Capacity-building providers can use these three actions to better incorporate learning into their programming[41]:

- Collect relevant data points, ensure entrepreneurs obtain value from reporting, and be relational. Three key indicators are revenue, employment, and investment.
- Support enterprises' collection and use of data to manage their performance and improve their businesses.
- Experiment with different approaches and improve delivery.

Villgro India is an example of a capacity-building service provider that incorporates learning into its delivery. Villgro is a social enterprise incubator that over time has learned it should focus not just on executing programs but also on prioritizing learning. To help it do this, Villgro India regularly collects data on the entrepreneurs it supports. It begins with a baseline evaluation when the enterprise joins its program and then holds periodic review meetings with the entrepreneurs, using data to measure progress.

against milestones and to assess the effectiveness of the support provided. This allows entrepreneurs to view regular performance assessments as adding value and has allowed Villgro to improve its programming model over time.

Lead by example

A last critical practice offered by highly effective capacity-building service providers is to lead by example, demonstrating to program participants what good practices look like in practice. This improves the likelihood that not only will entrepreneurs trust the advice service providers offer but also that service providers know what they are talking about from firsthand experience.

The reality is that many capacity-building providers are themselves growing businesses that face internal challenges. Good organization is a prerequisite to successful program delivery. Providing a good example can help contribute to better performance in service provision and greater likelihood of financial sustainability. This in turn helps ensure the continuity of the training programs and the impacts they generate.[42]

Capacity-building organizations can lead by example by taking the following actions:

- Have a clear and focused strategy to guide program delivery.
- Integrate a governance structure into your organization that includes successful entrepreneurs and diverse and inclusive perspectives.
- Build a well-organized, empowered, and capable team to enable your organization to deliver and grow.

- Diversify income streams to strengthen financial sustainability.

A lead-by-example capacity-building provider I have long admired is Village Capital (VilCap). I have watched VilCap evolve over the years as it has supported impact-driven start-up enterprises through corporate-sponsored accelerator programs and workshops, most often focused on specific sectors or industries, such as agriculture or fintech. VilCap consistently works to improve its own organizational performance. For example, several years ago VilCap used unrestricted donor funding to establish its back-office operation. This allowed it to integrate technology, implement new processes, and provide better support to acceleration programs. Many of the participant enterprises took note of this excellent back-office delivery, and VilCap wound up developing training as part of some of its programs on how others could leverage their own learnings.

Summary of key messages from this chapter

- The quality and effectiveness of the many types of capacity-building support offered along with capital vary significantly.
- After analyzing patterns revealed by eight years of programming support for more than 500 capacity-building programs, as well as an extensive literature review, five key considerations emerged that characterize cost-effective and impactful capacity-building programs, considerations codified with the acronym SCALE.
- *Selecting* the right enterprises to participate in capacity-building programs allows providers to identify the enterprises most motivated to grow, to tailor their programming to those enterprises' needs, and to form cohorts to best leverage peer-to-peer learning.
- *Charging* enterprises to participate in capacity-building program ensures that programming is delivered to participants motivated to improve. It also improves sustainability of program delivery over time.
- *Addressing* entrepreneurs' problems in capacity-building programs and maximizing learning from relevant peer experiences instead of teaching them what they ought to know is more likely to foster enterprise growth and productivity improvement.
- *Learning* to evaluate effectiveness by incorporating continuous self-examination and feedback loops into strategic planning and delivery of capacity-building programs helps providers to understand if their programming is working and how it can be honed to work more effectively.

- Become an *example*. Capacity-building organizations that lead by example and demonstrate how to implement organizational improvements can be more effective at training other enterprises to make those same changes, improving overall program effectiveness.

Notes

1. Bardasi E, Gassier M, Goldstein M et al. (2021) The Profits of Wisdom: The Impacts of a Business Support Program in Tanzania. *World Bank Economic Review* 35(2): 328–347. https://doi.org/10.1093/wber/lhz048. Accessed 17 Aug 2022.
2. Argidius Foundation (2021) How to Fulfill the Potential of Business Development Services Using SCALE. https://api.cofraholding.com/media/2612/report-fulfilling-the-potential-of-bds.pdf. Accessed 18 Aug 2022.
3. Roberts P, Lall S, Baird R et al (2016) *What's Working in Startup Acceleration: Insights from Fifteen Village Capital Programs*. Aspen Network of Development Entrepreneurs. https://www.andeglobal.org/publication/whats-working-in-startup-acceleration-insights-from-fifteen-village-capital-programs/. Accessed 18 Aug 2022.
4. Ferhman R (2020) *Decoding the ABCs of Effective Enterprise Acceleration: 10 Lessons from the Social Entrepreneurship Accelerator at Duke (SEAD)*. Social Entrepreneur Accelerator at Duke. https://centers.fuqua.duke.edu/case/wp-content/uploads/sites/7/2020/10/Decoding-the-ABCs-of-Effective-Enterprise-Acceleration_Oct-12.pdf. Accessed 19 Aug 2022.
5. Endeavor Insight (2018) Fostering Productive Entrepreneurship Communities: Key Lessons on Generating Jobs, Economic Growth, and Innovation. https://endeavor.org/wp-content/uploads/2021/09/Fostering-Productive-Entrepreneurship-Communities.pdf. Accessed 18 Aug 2022.
6. Kutzhanova N, Lyons T, Lichtenstein G (2009) Skill-Based Development of Entrepreneurs and the Role of Personal and Peer Group Coaching in Enterprise Development. *Economic Development Quarterly* 23(3):193–210. https://doi.org/10.1177/0891242409336547.
7. Cooney TM (2012) *Entrepreneurship Skills for Growth-Orientated Businesses*. OECD.
8. IPA (2016) Thinking Like an Entrepreneur: Boosting Small Business Growth with Mindset Training in Togo. https://www.poverty-action.org/study/thinking-entrepreneur-boosting-small-business-growth-with-mindset-training-togo. Accessed 19 Aug 2022.
9. IPA (2016) Thinking Like an Entrepreneur: Boosting Small Business Growth with Mindset Training in Togo. https://www.poverty-action.org/study/thinking-entrepreneur-boosting-small-business-growth-with-mindset-training-togo. Accessed 19 Aug 2022.

10. Azoulay P, Jones B, Kim J et al (2020) Age and High-Growth Entrepreneurship. *American Economic Review: Insights* 2(1): 65–82. https://www.aeaweb.org/articles?id=10.1257/aeri.20180582. Accessed 18 Aug 2022.
11. McKenzie D, Sansone D (2017) *Man vs. Machine in Predicting Successful Entrepreneurs: Evidence from a Business Plan Competition in Nigeria.* World Bank. http://hdl.handle.net/10986/29007. Accessed 18 Aug 2022.
12. See the case study in: Argidius Foundation (2021) How to Fulfill the Potential of Business Development Services Using SCALE. https://api.cofraholding.com/media/2612/report-fulfilling-the-potential-of-bds.pdf. Accessed 18 Aug 2022.
13. Argidius Foundation (2021) How to Fulfill the Potential of Business Development Services Using SCALE. https://api.cofraholding.com/media/2612/report-fulfilling-the-potential-of-bds.pdf. Accessed 18 Aug 2022.
14. Bardasi E, Gassier M, Goldstein M et al (2021) The Profits of Wisdom: The Impacts of a Business Support Program in Tanzania. *World Bank Economic Review* 35(2): 328–347.
15. This summarizes the responses from a May 2021 Dalberg survey of 36 organizations to the question "What are your reasons for not charging?".
16. Maffioli A, McKenzie D, Ubfal D (2020) *How Should Business Training be Priced? A Demand Experiment in Jamaica.* World Bank. http://hdl.handle.net/10986/34548. Accessed 18 Aug 2022.
17. Maffioli A, McKenzie D, Ubfal D (2020) *How Should Business Training be Priced? A Demand Experiment in Jamaica.* World Bank. http://hdl.handle.net/10986/34548. Accessed 18 Aug 2022.
18. Maffioli A, McKenzie D, Ubfal D (2020) *How Should Business Training Be Priced? A Demand Experiment in Jamaica.* World Bank. http://hdl.handle.net/10986/34548. Accessed 18 Aug 2022.
19. Bando R, Näslund-Hadley E, Gertler P (2019) *Effect of Inquiry and Problem Based Pedagogy on Learning: Evidence from 10 Field Experiments in Four Countries.* Interamerican Development Bank. https://www.nber.org/papers/w26280. Accessed 18 Aug 2022.
20. Bruhn M, Karlan D, Schoar A (2013) *The Impact of Consulting Services on Small and Medium Enterprises: Evidence from a Randomized Trial in Mexico.* World Bank Group.
21. Anderson S, McKenzie D (2020) *Improving Business Practices and the Boundary of the Entrepreneur: A Randomized Experiment Comparing Training, Consulting, Insourcing and Outsourcing.* World Bank. http://hdl.handle.net/10986/34979. Accessed 18 Aug 2022.
22. Anderson S, Chandy R, Zia B (2019) Pathways to Profits: Can Entrepreneurship Training Really Work for Small Businesses? *VoxDev.* https://voxdev.org/topic/firms-trade/pathways-profits-can-entrepreneurship-training-really-work-small-businesses. Accessed 18 Aug 2022.

23. Anderson S, Chandy R, Zia B (2019) Pathways to profits: Can entrepreneurship training really work for small businesses? *VoxDev*. https://voxdev.org/topic/firms-trade/pathways-profits-can-entrepreneurship-training-really-work-small-businesses. Accessed 18 Aug 2022.
24. Iacovone L, Maloney W, McKenzie D (2019) *Improving Management with Individual and Group-Based Consulting: Results from a Randomized Experiment in Colombia*. World Bank Group. http://hdl.handle.net/10986/31712. Accessed 18 Aug 2022.
25. Iacovone L, Maloney W, McKenzie D (2019) *Improving Management with Individual and Group-Based Consulting: Results from a Randomized Experiment in Colombia*. World Bank Group. http://hdl.handle.net/10986/31712. Accessed 18 Aug 2022.
26. Cai J, Szeidl A (2018) Interfirm Relationships and Business Performance. *Quarterly Journal of Economics* 133(3): 1229–1282. https://doi.org/10.1093/qje/qjx049. Accessed 18 Aug 2022.
27. Cai J, Szeidl A (2018) Interfirm Relationships and Business Performance. *Quarterly Journal of Economics* 133(3): 1229–1282. https://doi.org/10.1093/qje/qjx049. Accessed 18 Aug 2022.
28. Woodruff C (2018) *Addressing Constraints to Small and Growing Businesses*. International Growth Centre. https://www.andeglobal.org/publication/addressing-constraints-to-small-and-growing-businesses/. Accessed 18 Aug 2022.
29. Argidius Foundation (2019) Networking Works: Peer-to-Peer Business Networks Help Small and Growing Businesses Grow Revenues and Create Jobs. https://www.andeglobal.org/publication/networking-works-peer-to-peer-business-networks-help-small-and-growing-businesses-grow-revenues-and-create-jobs/. Accessed 18 Aug 2022.
30. Endeavor Insight (2018) Fostering Productive Entrepreneurship Communities: Key Lessons on Generating Jobs, Economic Growth, and Innovation. https://endeavor.org/wp-content/uploads/2021/09/Fostering-Productive-Entrepreneurship-Communities.pdf. Accessed 18 Aug 2022.
31. Endeavor Insight (2018) Fostering Productive Entrepreneurship Communities: Key Lessons on Generating Jobs, Economic Growth, and Innovation. https://endeavor.org/wp-content/uploads/2021/09/Fostering-Productive-Entrepreneurship-Communities.pdf. Accessed 18 Aug 2022.
32. Endeavor Insight (2018) Fostering Productive Entrepreneurship Communities: Key Lessons on Generating Jobs, Economic Growth, and Innovation. https://endeavor.org/wp-content/uploads/2021/09/Fostering-Productive-Entrepreneurship-Communities.pdf. Accessed 18 Aug 2022.
33. TechnoServe (2017) Accelerating Impact for Entrepreneurs: Lessons Learned from TechnoServe's Work with SGBs in Central America. https://www.technoserve.org/wp-content/uploads/2017/05/Accelerating_Impact_for_Entrepreneurs.pdf. Accessed 18 Aug 2022.

34. TechnoServe (2017) Accelerating Impact for Entrepreneurs: Lessons Learned from TechnoServe's Work with SGBs in Central America. https://www.technoserve.org/wp-content/uploads/2017/05/Accelerating_Impact_for_Entrepreneurs.pdf. Accessed 18 Aug 2022.
35. TechnoServe (2017) Accelerating Impact for Entrepreneurs: Lessons Learned from TechnoServe's Work with SGBs in Central America. https://www.technoserve.org/wp-content/uploads/2017/05/Accelerating_Impact_for_Entrepreneurs.pdf. Accessed 18 Aug 2022.
36. Åstebro T (2021) Impact Measurement Based on Repeated Randomized Control Trials: The Case of a Training Program to Encourage Social Entrepreneurship. *Strategic Entrepreneurship Journal* 15(2): 254–278. https://doi.org/10.1002/sej.1391. Accessed 18 Aug 2022.
37. Anderson S, Chandy R, Zia B (2019) Pathways to Profits: Can Entrepreneurship Training Really Work for Small Businesses? *VoxDev*. https://voxdev.org/topic/firms-trade/pathways-profits-can-entrepreneurship-training-really-work-small-businesses. Accessed 18 Aug 2022.
38. Drexler A, Fischer G, Schoar A (2014) Keeping It Simple: Financial Literacy and Rules of Thumb. *American Economic Journal: Applied Economics* 6(2): 1–31. https://www.aeaweb.org/articles?id=10.1257/app.6.2.1. Accessed 18 Aug 2022.
39. Arráiz I, Bhanot S, Calero C (2019) *Less Is More: Experimental Evidence on Heuristic-Based Business Training in Ecuador*. Interamerican Development Bank. https://www.idbinvest.org/en/publications/report-less-more-experimental-evidence-heuristics-based-business-training-ecuador. Accessed 21 Aug 2022.
40. Global Accelerator Learning Initiative. (2021) A Rocket or a Runway? Examining Venture Growth During and After Acceleration. https://www.galidata.org/publications/a-rocket-or-a-runway/. Accessed 18 Aug 2022.
41. Global Accelerator Learning Initiative. (2021) A Rocket or a Runway? Examining Venture Growth During and After Acceleration. https://www.galidata.org/publications/a-rocket-or-a-runway/. Accessed 18 Aug 2022.
42. Global Accelerator Learning Initiative. (2021) A Rocket or a Runway? Examining Venture Growth During and After Acceleration. https://www.galidata.org/publications/a-rocket-or-a-runway/. Accessed 18 Aug 2022.

19

"Scale Based on Evidence of Effectiveness"—An Interview with Nicholas Colloff

This chapter summarizes my interview with Nicholas Colloff, Director at the Argidius Foundation a leading philanthropic funder of enterprise development to tackle poverty in emerging markets.

Question 1: How did you first get involved in the design and delivery of capacity building for small and growing businesses?

Nicholas: I was building a micro- and small-enterprise bank in what is now the Republic of North Macedonia. It was a post-communist space with many emerging micro and small businesses. We recognized that networking businesses together was likely to improve their survival and resilience rate, so we built alongside the bank a business network both for our clients and for nonclients. Through that network, we provided basic capacity-building services on issues such as financial management, sales and marketing, human-resource management, and to a certain extent, imports and exports. That network grew alongside the bank for several years and was then folded into a wider project.

Recognition of the need for capacity building came after success in doing something very practical. We built an institution that worked, was valued, and seemed to have a positive net result on the resilience and growth of

K. Hornberger, *Scaling Impact*,
https://doi.org/10.1007/978-3-031-22614-4_19

businesses. That was my first taste of the nonfinancial part of enterprise development.

Question 2: Tell us about your current role. What is Argidius, and why has it chosen enterprise development as its focus area?

Nicholas: Argidius is a private family foundation. It's always been focused on enterprise solutions to poverty. It has addressed those solutions in a number of ways, based mostly on helping to develop financial inclusion through microfinance. But one recognition that came out of that work in financial inclusion was that if you want to create genuine change in the dynamics of poverty, the best proxy indicator is full-time job creation in the formal economy.

What you want to do is create small or medium productivity-driven enterprises (SMEs) that sit within the formal sector. If you look at developed economies, such as Germany or Switzerland, around 40 percent of GDP is composed of SMEs. If you look at an emerging economy, like Kenya, SMEs represent about 15 to 20 percent of GDP.

Argidius decided to focus on SMEs in the formal sector to help create formal employment in productive enterprises. You do that by taking one of two broad tracks. You could determine what nonfinancial capacity building these enterprises need, or you could look at access to finance. We decided to focus principally on the nonfinancial support SMEs need to grow.

The second question became, then, do these programs actually help businesses grow? When I arrived, a report was circulating that said business accelerators are critical to growth. However, I had two hesitations about this approach. First, we didn't know whether the business accelerators actually had an impact. It was possible they just chose businesses already making progress, that is, businesses that would have grown either way. Second, many kinds of interventions didn't fit the accelerator pattern. Interventions such as having business networks, individual consultancies, or coaching and mentoring also seemed to have potential to help businesses become more resilient and grow.

So we decided to find out what types of interventions work and what characteristics make them work more effectively. We used our data to measure the incremental revenue, employment growth, and capital raised across interventions. We chose interventions that we thought looked good, at least on paper. Then we disaggregated our data over time to see which of the interventions consistently secured the greatest incremental revenue growth. Finally, we brought all of this evidence together to see if there was a pattern. This information allows people to focus their efforts on things that work, and hopefully funders will be better able to target their money toward things that work too.

Question 3: How do you define capacity-building services for enterprise growth?

Nicholas: Capacity building for us is any intervention that increases the knowledge that the entrepreneur or enterprise has at its disposal to successfully run and manage the business going forward. That could mean very generic things, such as sales, marketing, or financial management support. It could also involve technical interventions, such as how to reduce waste in a mango business or how to attract and retain management teams. So the definition is fairly broad, but it comes down to any structured intervention that helps the business gain additional knowledge that helps it become resilient and grow.

Question 4: What are the differences in the types of capacity-building support a microfinance institution would provide versus the type of support a venture capital firm might provide?

Nicholas: The underlying similarity is that capacity building transfers knowledge that can be intelligently applied in a way that changes business practices. However, businesses at different stages of development have different levels of sophistication. If you're thinking about formalizing a business, you need to know the business basics. You must answer questions such as do I have the right things to sell, am I selling the right way, how do I separate my finances as a person from the finances of the business? At the other end of the spectrum, you have much more sophisticated businesses asking questions such as do I understand the market, does the market exist or am I trying to create a market, what does it mean to think about cash flow in relation to my attractiveness as an investment proposition?

The level of what you must pay attention to and the complexity of the issues both increase as the business develops. You'll need more knowledge and more sophisticated arrangements of your business practices, and you will need to understand how all these things fit together to create and maintain a business that's scaling and growing.

Question 5: Some private investors believe they don't need to provide nonfinancial services because they add costs and because entrepreneurs know best how to run their business anyway. Why do you think it is critical that capital is combined with capacity building?

Nicholas: If you actually look at the successful growth patterns followed by entrepreneurs who have built large-scale corporations, you'll find they are made up of a whole range of knowledge providers, both formal and informal. The entrepreneur may have had a mentor who provided relatively sophisticated consulting advice. Many entrepreneurs sit in teams and often get support to help select those teams. The most successful entrepreneurs have experience working in Fortune 100 companies, so they understand how to work in organizations that are building capacity. If you think about innovation, we often have this image of the lone innovator. But when you look at the actual story, lone innovators turn out to be sitting within ecosystems that enable them to innovate and be successful.

If you're an entrepreneur in an emerging market economy, you probably haven't had that experience base. The closer you look into such markets, the more you recognize that people there are sitting in informal ecosystems and that privilege people with certain backgrounds. So, if you want to extend opportunities for growth, you have to do it more consciously, and that's what business capacity building does. If you don't create the structures and networks, the result is less resilient businesses that grow less quickly.

The COVID-19 pandemic has been an experiment in taking the capacity-building approach. The enterprises that received such support were much more resilient than those who didn't. Our portfolio companies created net employment gain when comparable businesses were shedding revenues and employment.

So if you're an investor who wants to hit more home runs, you should find businesses that have built that kind of support around them. Often that support is built through capacity-support providers.

Question 6: What are the different types of capacity-building services?

Nicholas: We put them into a few wide buckets:

- Business incubators and accelerators pass a cohort of enterprises through a relatively structured program.
- Business networks gather resources together.
- Individual consultants offer coaching or mentoring.
- Professional service consultancies provide significant capacity inputs as well as ensuring the business meets its regulatory requirements for audits and so on.

What we found over our eight years of exploring almost 500 discrete cohorts in 100 different programs across 50 different organizations is that it didn't really matter what a service called itself. To be effective, what mattered more was how capacity building was provided and the structure it used.

On the training side, the broad categories most entrepreneurs want to address are:

- Underlying business knowledge
- Sales and marketing capabilities
- Talent-management capabilities
- Financial-management capabilities
- Governance
- Ability to access appropriate markets.

Underlying these categories is what I call planning, which isn't just about the business plan. It's about entrepreneurs continually scanning their horizon and identifying what's next, about being consistently in planning mode. We can help them develop approaches to maintaining that posture.

Additionally, soft skills flow through all or many of those skills. As an entrepreneur, it is important to recognize your strengths and to understand how to create and lead a team whose members bring different strengths. During the COVID-19 pandemic we also discovered the importance of psychological resilience. Being an entrepreneur can be a challenge. It is often lonely and exhausting. The soft skills needed include taking care of those feelings intelligently so you don't burn out.

Question 7: What have you learned about the types of capacity-building services that are most effective?

Nicholas: The first step is to help the entrepreneur identify its current most significant challenge and then to help resolve it. For example, solving a cash-flow problem opens a window for the entrepreneur to learn about cash flow and to establish the right processes so this problem doesn't arise again. By helping entrepreneurs identify a few of the key problems they need to resolve allows them to take those learnings and to work successfully to prevent those problems from happening again.

Many programs that fail do so because they assume that people enjoy learning for the sake of learning. But if cash flow is the most important thing on an entrepreneur's mind, it presents the opportunity for the service provider to help solve that problem and to make it a prompt from which the entrepreneur can learn. This is called a reverse curriculum, because it's the opposite of what you did in school, where the teachers tell you how to do something and then send you off with homework to practice solving the problem. Our approach here is the reverse of that. *First* we solve the problem, and *then* we encourage the entrepreneur to go back and learn how to avoid having that problem again. That's the most powerful approach to learning.

We jokingly describe the process by asking people, "When you buy a new washing machine, do you read the manual before turning it on?" Nine out of ten people don't. Only when some strange light flashes at them do they turn to the manual.

The second most powerful element is for capacity builders to know what kind of business they are dealing with and the stage of business they are best quipped to help. Is it a formalizing business or a larger business? They must get that right and stay focused. Then they must help the business understand whether it is selecting the kind of capacity builder it needs. Some of our partners are effective at working with larger businesses facing problems that don't really emerge in smaller businesses, and vice versa. Good interveners are clear about their work in this space.

Another effective strategy is peer-to-peer learning. You learn better when you feel like you're learning from peer companies of roughly the same age, the same size, but not necessarily in the same field. Among peers, similar problems emerge, and the solutions will be similar. This kind of peer-to-peer learning is really important and can be cost effective.

In a famous example from China, businesspeople were brought together and after the first facilitated conversation, they were left to meet to chat once a month on their own for about six to eight months. Out of that came new business contacts, and participants learned from one another. Those businesses grew 10 percent faster than businesses that didn't have the opportunity for those chats and that engagement.

We've also found that charging is important, not necessarily the full cost but a meaningful cost. For the business, charging means that it chooses to work with the service provider and to pay money for the service. That's likely to lead to greater commitment from the business, and it's likely to make the provider look at the business as more of a customer or client rather than as a beneficiary.

We had a partner in Central America that changed just two things about its program. They charged for the first time, and they made the program slightly less intense because they had determined they had been asking people to learn and do too many changes at once. They got fewer applicants after they started charging, but the quality of applicants improved. The newer applicants were much more committed and engaged and saw the program as something they genuinely wanted to do. The impact of these program changes improved results by a multiple of three, a level that has been consistently achieved in subsequent cohorts.

Question 8: What are some examples of organizations that perform highly effective capacity-building services?

Nicholas: One organization we've worked with for a long time is TechnoServe, a US-based NGO. We worked with them to build a particular business acceleration program in Latin America that is now on its third iteration. In each of these three iterations, over 800 businesses have passed through the program. TechnoServe has been able to take that experience and build an entrepreneurship practice.

They've very effective and deliberate about it. They used one measure, return on total investment, and over time they have been able to bring the quality of their programs up each year. The businesses they work with are what we call dynamic businesses, that is, businesses with revenues between $50,000 and $400,000, and they are usually family owned. The businesses often reach out for support when they come to some crunch point, either because of a generational shift in the business or something has happened in the market that requires them to think differently about their business.

At the other end of the spectrum is Villgro Africa, which works with health-related businesses. Interestingly, they provide both capacity support and seed funding. They've been able to demonstrate a very serious process of careful mentoring of businesses, helping the businesses to think through effective business models and providing very early, relatively small amounts of financial support. Their approach has helped to accelerate the growth of a number of very interesting businesses.

Another good example is the African Management Institute, which focuses on business practices. They have created a diagnostic that enables early-stage businesses to think deeply about the two or three practices they need to work on if their businesses are going to change and grow. African Management Institute can then tailor its intervention to focus on those practices and harness the peer-learning effect as well.

Another good example is PUM, the Dutch experts' group, which provides very senior technical advising, often looking at firm processes. This examination can lead to hugely significant differences, often by streamlining processes, cutting out costs, and fostering thinking about adding significant levels of value to businesses.

One example of this intervention was with a Nicaraguan nut business that sold its nuts, no added value at all, into Mexico. Then Americans came and dumped their nuts into the Mexican market, and left the Nicaraguan business with no market to sell to. PUM turned the business around by thinking about how to add value to the nuts. In terms of volume, the business shrunk, but value multiplied many times over, and the business is now a profitable nut business with a range of meaningful products sold in Germany, the Netherlands, and France. This was a relatively inexpensive change, achieved in only a few months, but it made a huge difference.

Question 9: What are some of the trends and challenges in scaling funding for capacity-building services?

Nicholas: The challenge is the sustainability of the institutions. Though the funding has slowly improved it's still a constant challenge to find funding.

Charging entrepreneurs can help, but not to the point of creating sustainable institutions, because you don't have the ability to charge the full economic cost of what you're doing. In developed countries, 94 percent of all accelerator programs are subsidized in some way. They don't function by capturing the incremental revenue of the underlying business. The funding comes from public sources, universities, corporates, or philanthropy. Most countries decide that business development services are important and should be provided like a blended public good. You also see that in emerging countries like Guatemala, where the government provides funding to regional business development centers.

The challenge is to consistently make a better business case for the importance of capacity-building services. As countries move from low-income to middle-income to higher-income economies, the level of government support provided is developed, corporate support is developed, and university institutions want to do business incubation or acceleration as part of their offerings to the community.

Funding is often "stop and go," so organizations need to be able to effectively expand and contract. CEED, for example, is good at providing a core offering that can be funded by the entrepreneurs, who can pay for it because CEED works with somewhat larger dynamic and venture businesses. CEED also runs programs that can be retracted if funding disappears. Some of its staff are consultants, as opposed to paid permanent staff, so it can add people to deliver programs but also contract programs if money is not available. TechnoServe does something similar. This is not ideal, but it allows providers to deal with fluctuating funding.

But the broad trend is that as countries hit a middle-income status, you find local and international corporates interested in the space and governments willing to fund services, sometimes through their back donors. Kenya, for example, has a World Bank loan of $150 million, which is being redesigned to help local economies build their capacity. Everyone is interested in identifying high-growth firms in Kenya to create formal employment. Each year 700,000 people enter the Kenyan workforce, but only 55,000 enter formal employment, and a quarter of those formal jobs have been created by enterprises supported by business development services. This supports the position that putting in place a more coherent ecosystem for the capacity-building services will generate the formal employment needed to move from low- to middle-income status.

In addition, capacity building becomes attractive because it also ensures a degree of social stability. Look at Uganda, for example, where almost half of the population is under 15. Capacity building can help provide the jobs that will enable Uganda's young people to move out of poverty and have meaningful lives.

Question 10: What would you like other donors and funders to do differently with regard to supporting capacity-building providers?

Nicholas: I think the best possible shift would be twofold. First, the whole purpose should be to scale and get people to think about that intelligently. Second, donors and funders should think about funding organizations over time, not just about funding programs. This wouldn't necessarily mean spending more money, just changing the mindset.

Often funders, particularly some of the larger institutional funders, will say they want to fund a successful business development ecosystem. But when you look at how they fund their capacity-building services, they're funding programs but not the full cost of those programs or the organizations behind them. They then wonder why the organizations don't make progress toward becoming more robust and effective. It's because they're not being paid properly to do what they are intended to do.

So, those steps would constitute the best possible shift. And sometimes it makes a difference to think, what's the genuine cost of delivering this service to entrepreneurs? Are we paying this organization the genuine cost? Or are we expecting others in private finance, in addition to ourselves, to subsidize them? The natural conclusion from asking this is that I wouldn't see any reason to subsidize a development finance institution or bilateral donor to do what it should already be doing more effectively.

20

Conclusion—What We Can Do to Scale Impact

While global challenges continue and the daily grind can distract us from what is truly important, I hope this book has shown you that real progress is already happening and that greater progress is possible. We can work within the capitalist system to reimagine rather than revolutionize it. I have suggested six concrete paradigm shifts in how we collectively can think differently about providing finance and investment. Now more than ever as public and private investment in emerging markets hits new heights, we need to shift paradigms to build and scale the impact economy toward achieving the SDGs.

I have long believed that value creation is in fact impact maximization. If finance providers of all types are truly maximizing value for their stakeholders, they are scaling impact. They are allowing growth by putting capital to work, and they are encouraging impact by intentionally valuing the use of money toward ends that help both people and the planet.

Shifting paradigms toward scaling impact will not only be about changing beliefs or creating new tools such as blended finance, innovative finance, and impact investing; it will also be about changing our actual behavior. More importantly, it will be about changing the status quo practices of the ecosystem of actors involved in deploying capital. We need to shift toward impact, whether made by the small microfinance institution providing financial services to individuals or large pension funds investing in diversified growth-oriented asset managers—everyone must take impact more seriously. Below I offer concrete suggestions by type of actor based on my experience and the learnings from this book.

K. Hornberger, *Scaling Impact*,
https://doi.org/10.1007/978-3-031-22614-4_20

Financial service providers can help individuals achieve financial health by listening to client needs and using that knowledge to better segment and tailor their products and services to truly address those needs. They can do this in a way that ensures consumer protections, including our right to privacy over our personal data and transparency concerning fair pricing and profit. They can also seek ways to offer more holistic bundled services that include options like providing nonfinancial capacity-building support where helpful. Moving beyond traditional financial access through savings and loans, providers can offer broader complementary services that help customers grow and achieve financial health. In so doing, they will help extend economic security to a much larger portion of the populace, moving a long way toward ending poverty.

Early-stage investors can help small businesses grow by offering patient capital that accepts risk and has more flexible time horizons to return capital. This starts by recognizing and responding to the diverse needs of enterprises. Livelihood-sustaining, niche, dynamic, or high-growth enterprises each need very different financial and accompaniment solutions. More specifically, much more attention needs to be given to the large and growing financing needs of non-high-growth ventures that inhabit and service most of our daily lives across the globe. This also means customizing financing to the needs of small businesses. Alternative financing, often referred to as "mezzanine" finance or "structured exit," is an approach that sits between commonly deployed pure debt and pure equity solutions. These contracts take the best parts of equity (accompaniment, flexibility, partnerships, and patience) and mix them with the best parts of debt (self-liquidation and cost), without pressuring founders to achieve exponential growth and force the eventual sale of their business to satisfy investors. Using these alternative products more effectively can go a long way toward closing the large financing gaps at the heart of many social and environmental challenges.

Growth-stage investors can prioritize impact to deliver greater value to a diverse set of stakeholders. They can do this by going beyond ESG to incorporate clear positive intent, ensuring clear logic guides contributions of investment capital, linking it to outcomes and objectively measuring and reporting on the impact of investments made. They can also move beyond making this a theoretical exercise by setting up dedicated impact funds, integrating financial and impact metrics, developing and using impact rating tools to screen investments, and evaluating and reinforcing impact goals at investment exit. They can further ensure impact by choosing to provide assurance to external parties by having their impact management and measurement practices audited by external impact auditors.

Development finance institutions (DFIs) can become more catalytic by expanding the use of their blended finance windows to accelerate innovation and market development. DFIs can do this by providing first-loss capital, such as subordinated debt or junior equity, or even direct grants into the capital stack or incentive payments to boost return potential for other investors. Or they can provide design-stage funding, guarantees, or technical assistance facility funding to de-risk the investment opportunity. But more importantly DFIs should not shy away from using their blended finance windows and their ability to generate not only financial but also ecosystem additionality, such as by demonstrating effects, building new markets, or setting standards in a new area of business activity. Such efforts should be at the core of the existence and continued public support of DFIs.

Donors and philanthropic organizations can proactively base funding on results achieved by encouraging the greater use and adoption of innovative finance tools such as impact bonds or social success incentive notes. To see these tools mainstream will require increased commitment from donors and philanthropic funders to experiment, learn, and share evidence with a broader community. The larger outcome fund of funds is a first positive step in that direction as we are still in the early stages of seeing these approaches reach their potential.

As a closing thought, while I am an optimist and dare to see the world as it could be and to try to build it, I don't think you have to be an optimist to see that these important changes can tilt our financial system toward impact. The approaches outlined here are rooted in pragmatism as well as optimism. We need to adapt to survive the impending consequences of climate change, growing inequalities, the spread of authoritarianism, and the many other threats to human existence and flourishing. By more intentionally placing impact as a goal of our practices in providing finance and investment, we can keep the climate bearable for everyone, solve poverty for hundreds of millions of people, and ensure that all people everywhere lead lives with dignity and opportunity. I hope you will join me in the endeavor to scale impact and make finance and investment work for a better world.

Glossary

Accelerator — Time-limited programs that work with cohorts or "classes" of ventures to provide mentorship and training, with a special emphasis on connecting early-stage ventures with investment

Alpha — A term used in investing to describe an investment strategy's ability to beat the market

Angel investor — Individuals or networks with resources who invest in seed stage or very early-stage start-ups (typically in exchange for equity) and provide additional support (often in the form of expertise, coaching, or mentoring)

Asset class — A grouping of financial instruments that behave similarly in the market

Assets — Physical or intangible resources such as building, equipment, and brands that are expected to generate value

Assets under management (AUM) — The total market value of the investments that a person or entity manages on behalf of clients

Auctions — A type of results-based financing where funders commit to funding upfront, provide initial funding to implementing partners to get started or pilot, and then award full funding or a share of funding via auction to the most competitive bidder, based on outcomes achieved and cost efficiency

Awards or prizes — Competitions that help surface solutions, partners, and ideas to help support innovation and implementation

Base of the pyramid (BOP) — Individuals at the bottom fifth of the economic pyramid of income generation (e.g., often those considered earning less than $5.50 per day, depending on context)

Blended finance — The mobilization of private capital by creating financial structures that allow impact-oriented donors and commercial capital providers to

K. Hornberger, *Scaling Impact*,
https://doi.org/10.1007/978-3-031-22614-4

deploy capital alongside each other and achieve goals that would not have been possible otherwise.

Business development services — Support for the growth of micro, small, and medium-sized enterprises (MSMEs) through training, technical assistance, marketing assistance, improved production technologies, and other related services

Capital structure — Also known as capital stack or waterfall, is made up of various types of financing used for the business operations. This includes both external sources of debt and equity as well as internal sources, which is earned in the form of net profits or retained earnings

Capacity-building services — Also known as enterprise support, business development services, or technical assistance (TA), are nonfinancial services used to strengthen or enhance enterprise performance

Catalytic capital — Investment capital that is patient, risk-tolerant, concessionary, and flexible in ways that differ from conventional investment. It is an essential tool to bridge capital gaps and achieve breadth and depth of impact, while complementing conventional investing

Challenge funds — These funds galvanize people outside the funding organization to develop innovative solutions to development challenges. The funding organization awards the prize funds to the organization(s) with the best solution that achieves desired outcomes and helps surface new types of partnerships

Commercial capital — Investment capital that expects to be invested and returned at or above market rates

Concessional capital — Investment capital or funding that expects to be invested and returned at below-market rates

Convertible note — A fixed-income corporate debt security that yields interest payments but can be converted into a predetermined number of common stock shares. The conversion from note to equity can be done at certain times during the note's life and is usually at the discretion of the noteholder. As a hybrid security, the price of a convertible note is especially sensitive to changes in interest rates, the price of the underlying asset, and the issuer's credit rating

Customized financing products — This approach adapts financial products to enterprise-specific needs and local market contexts. Highly customized "mezzanine" financial products are especially useful where straight equity or debt investments are difficult or do not match enterprise needs

Debt financing — A fund raising strategy where an individual or enterprise sells a debt instrument to individual and/or institutional investors. In return for lending the money, the individuals or institutions become creditors and receive a promise that the principal and interest on the debt will be repaid

Design funding — A type of blended finance where the transaction design or preparation is grant funded.

Development finance institution (DFI) — Specialized national or multinational banks or subsidiaries setup to support private sector development in developing countries

Development Impact Bond (DIB) — A results-based contract in which one or more investors provide working capital for social programs, service providers (e.g., implementors) implement the program, and one or more outcome funders pay back the investors their principal plus a return if, and only if, these programs succeed in delivering results. In a DIB, the outcome payer is typically a private donor or aid agency

Digitalization — The use of digital technologies to change a business model and provide new revenue and value-producing opportunities; it is the process of moving to a digital business

Donor-Advised Fund (DAF) — A tax-preferred philanthropic vehicle like a private foundation. A donor can establish a DAF with an initial tax-deductible contribution, and then recommend the DAF donate funds to other nonprofits later. This allows donors to separate their timing of the tax decision from the giving decision, and to give money out over time while claiming a tax benefit in the year (or years) most beneficial to them

Dynamic Enterprise — SGBs that operate in established "bread and butter" industries (such as trading, manufacturing, retail, and services) and deploy proven business models. Many are well-established and medium-sized, having steadily expanded over a number of years. They seek to grow by increasing market share, reaching new customers and markets, and making incremental innovations and efficiency improvements—but their rate of growth is typically moderate and tempered by the dynamics of mature, competitive industries

EBITDA — A widely used measure of corporate profitability, which stands for Earnings Before Interest, Taxes, Depreciation, and Amortization

Employee Stock Ownership Plan (ESOP) — An employee ownership model in the United States that allows a company to transfer full or partial ownership to employees

Endowment — Donation of money or property to a nonprofit organization, which uses the resulting investment income for a specific purpose

Enterprise support services — See capacity-building services

Environment, Social, and Governance (ESG) — A set of standards for a company's behavior used by socially conscious investors to screen potential investments

Evergreen fund — A fund that does not have a limited life, rather it exists in perpetuity. Funders in evergreen funds will need to have clauses in their funding contracts that specify how they will get their capital out of the fund. This will generally be through a combination of liquidity events such as dividends and selling their shares in the fund to other funders

Exchange traded funds (ETFs) — A type of pooled investment security that operates much like a mutual fund. Typically, ETFs will track a particular index, sector, commodity, or other assets, but unlike mutual funds, ETFs can be purchased or sold on a stock exchange the same way that a regular stock can

Family office — Private wealth management advisory firms that serve ultra-high-net-worth investors

Financial health — Financial health is achieved when an individual's daily systems help build the financial resilience to weather shocks, and the ability to pursue financial goals

Financial inclusion — Individuals and businesses have access to useful and affordable financial products and services that meet their needs (transactions, payments, savings, credit, and insurance) delivered in a responsible and sustainable way

Financial intermediary — An entity that acts as the middleman between two parties in a financial transaction

Financial service provider (FSP) — Organizations that provide banking, loans, money transfers, and financial options to customers

Fintech — Innovative technologies integrated into financial services to improve and automate the delivery and use of financial services

First-loss capital — A type of blended finance where public or philanthropic investors are concessional within the capital structure

Flexible capital — Capital which has unspecified use and can be used for anything the recipient wishes to use it for

Graduation approach — A holistic livelihoods program designed to address the multi-dimensional needs of extreme poor households. It consists of five core components: time-limited consumption support; a savings component; an asset transfer; training in how to use the asset; and life skills coaching and mentoring. The theory of change underlying the model is that this mix of interventions, offered in the appropriate sequence, will help the ultra-poor to "graduate" out of extreme poverty within a defined time period

Grace period — A period which allows a borrower to delay payment for a short period of time

Grant financing — Fundraising by an individual or enterprise that does not expect to be returned

Growth capital — Capital spent on hiring people, investing in new product development, putting systems into place, and/or marketing anything that helps build and scale business into the future

Guarantee — A type of blended finance that is designed to protect investors from incurring losses as the result of an investment opportunity that carries a high degree of risk

High-growth Venture — SGBs that pursue disruptive business models and target large addressable markets. These enterprises have high growth and scale potential and tend to feature the strong leadership and talent needed to manage a scalable business that pioneers completely new products, services, and business models

Impact investing — An investment strategy that seeks to contribute solutions and measure progress towards achievement of social and environmental goals

Impact management platform (IMP) — A collaboration between leading providers of public good standards, frameworks, tools, and guidance for managing sustainability impacts

Impact measurement and management (IMM) — The process of identifying the positive and negative effects of a business' activities on people and the planet, and managing these effects toward the business and/or the investor's social or environmental objectives

Impact thesis — A succinct and evidence-based proposition that indicates how an investment strategy will achieve its intended social or environmental impact

Impact-linked finance — Refers to linking financial rewards for market-based organizations to the achievements of positive social or environmental outcomes

Incentive payment — Where a donor funds a program that gives direct grants or incentives to investment funds or financial institutions that serve specific segments or meeting predefined objectives

Incubator — Institution that helps ventures define and build their initial products, identify promising customer segments, and secure resources

Inequality — The quality of being unequal or uneven, such as social disparity or disparity of distribution of access to opportunity

Investment thesis — Lays out the types of companies or projects that a funder invests in and how they create value. The thesis should identify the sectors, geographies, and stage of development they seek to fund as well as the type of instrument and parameters of investment they seek to place

Junior (subordinated) debt — Unsecured and has lower probability of being paid back should an investee default, since more senior debt is given priority

Limited Partners (LPs) — An investor into a fund or business that does not have a day-to-day role. Most venture capital and private equity funds are structured so that their investors are LPs. A fund is the General Partner (GP) that manage assets on behalf of its LPs

Livelihood-sustaining Enterprise — Small businesses selling traditional products and services. These businesses may be either formal or ready to formalize. They tend to operate on a small scale to serve local markets or value chains, often in sectors such as retail and services, and deploy well-established business models

Mezzanine financing — Hybrid financing that combines elements of debt and equity to create funding that is more flexible than pure debt or equity

Microfinance — A type of banking service provided to unemployed or low-income individuals or groups who otherwise would have no other access to financial services

Millennials — Individuals born in the 1980s or 1990s

Missing middle — In terms of business, this refers to a company that is too small for big investors and too big for small investors

Mobile money — A service in which the mobile phone is used to access financial services

Niche Venture — SGBs that create innovative products and services but target niche markets or customer segments. They seek to grow but often prioritize goals other than massive scale—such as solving a specific social or environmental problem, serving a specific customer segment or local community, or offering a product or service that is particularly unique or bespoke

Non-bank financial institution (NBFI) — Institutions that provide certain types of banking services but do not have a full banking license (e.g., savings and loan cooperatives, CDFI, fintechs, etc.)

Nonfinancial services — Support to an individual or enterprise that is not financial in nature. This can include knowledge or skills transfer and other types of capacity-building services.

Official development assistance (ODA) — Government aid that promotes and specifically targets the economic development and welfare of developing countries

Off-grid energy (OGE) — Being off the grid means not being physically hooked up to utilities by wires, pipes, or cables. Off-grid homes therefore rely completely on their own energy sources, which can often be renewable energy sources such as the sun and the wind

Open-ended capital vehicle — This approach to fund management extends the time horizon using evergreen or open-ended legal forms, rather than traditional closed-end funds. It responds to the longer time horizons often required to create and capture value required for investments in agriculture, education, health, and other social services in emerging markets

Operating principles for impact management (OPIM) — A reference point against which the impact management systems of funds and institutions may be assessed. They draw on emerging best practices from a range of asset managers, asset owners, asset allocators, and development finance institutions

Outcomes — The goals and objectives a company aims to achieve which come as a direct result of outputs

Outcomes-based contract — Bilateral agreements between a payor and service providers. Under the arrangement, service providers receive some funding from the payor to operate the program and receive reimbursement for full project/program costs and/or additional performance payments if they achieve agreed-upon outcomes

Outcomes-based financing — A financing contract where the funder only pays once the pre-agreed social and/or environmental outcomes have been achieved by the service provider. Like results-based financing

Outcomes-based pricing — Bilateral agreements between an investor and service providers. Under the arrangement, service providers receive working capital from the investor to implement a project, and the pricing of interest rate for repayments is lower or higher depending on achieving the outcomes agreed-upon during project implementation

Patient capital — An approach to enterprise financing that understands and recognizes the diverse needs of enterprises, prioritizes impact, has high risk tolerance uncorrelated with financial rewards, and maintains a more flexible time horizon to return capital than traditional closed-end venture or PE funds offer

Pay-for-success (PFS) — A set of innovative outcomes-based financing and funding tools that directly and measurably improve lives by driving resources toward results. These tools center on four core principles: clearly defined outcomes, data-driven decision-making, outcomes-based payment, and strong governance and accountability

Poverty — A state or condition in which a person or community lacks the financial resources and essentials for a minimum standard of living. Poverty means that the income level from employment is so low that basic human needs can't be met. Poverty-stricken people and families might go without proper housing, clean water, healthy food, and medical attention

Poverty line — Based on information about basic needs collected from 15 low-income countries, the World Bank defines the extreme poor as those living on less than $2.15 a day using 2017 purchasing power parity (PPP). In addition, the World Bank now also reports on two higher-value poverty lines: $3.20 and $5.50 per day. These lines, which are typical of standards among lower and upper-middle-income countries, respectively, are designed to complement, not replace, the $2.15 international poverty line

Preferred stock — A type of share that does not confer voting rights but is paid out before common stock in the event of liquidation. Preferred stockholders usually have no or limited voting rights regarding corporate governance. In the event of liquidation, preferred stockholders' claim on assets is greater than common stockholders but less than bondholders. As a result, preferred stock has characteristics of both bonds and common stock, which enhances its appeal to certain investors

Private capital — The umbrella term for investment, typically through funds, in assets not available on public markets. It includes private equity, venture capital, private debt, real estate, infrastructure, and natural resources.

Private equity — A type of alternative investment in which the investors purchase shares in privately held businesses.

Project preparation facility — An entity setup with the aim to strengthen and shorten the project preparation stage, facilitating loan approval and project execution mainly of infrastructure projects.

Redeemable shares — Shares owned by an individual or entity that are required to be redeemed for cash or for another such property at a stated time or following a specific event. Essentially, they are ownership shares with a built-in call option that will be exercised by the issuer at a predetermined point in the future.

Responsible exit — The practice of not just maximizing the returns of an investment upon exit but also of ensuring that the enterprise is set up for sustained impact through operations that benefit all stakeholders, rather than only shareholders

Results-based financing (RBF) — An umbrella term referring to any program or intervention that provides funding to individuals or institutions after agreed-upon results are achieved and verified

Revenue- or royalty-based lending — Raising capital from investors who receive a percentage of the enterprise's ongoing gross revenues in exchange for the money invested. In a revenue-based financing investment, investors receive a regular share of the income until a predetermined amount has been paid. Typically, this predetermined amount is a multiple of the principal investment. Unlike pure debt, however, no interest is paid, and there are no fixed payments

Risk mitigation — A strategy in blended finance to prepare for and lessen the effects of threats faced by a business

Segmentation — The process by which you divide your customers or dataset into segments up based on common characteristics

Service delivery models (SDM) — Supply chain structures, which provide services such as training, access to inputs, and finance to farmers

Shareholder — A person, company, or institution that owns at least one share of a company's stock. It is essentially someone who owns part of a company with certain rights and responsibilities in return for ownership

Simple Agreement for Future Equity (SAFE) — A financing contract between an investor and an entrepreneur that provides rights to the investor for future equity in the company, similar to a warrant, except without determining a specific price per share at the time of the initial investment. The SAFE investor receives the future shares when a priced round of investment or a liquidity event occurs. SAFEs are intended to provide startups seeking initial funding with a mechanism similar than convertible notes and to reward startup investors with discounts on future shares

Small and growing business (SGB) — Defined as commercially viable businesses with five to 250 employees that have significant potential and ambition for growth. Typically, SGBs seek growth capital from USD $20,000 to $2 million

Social enterprise — Defined as enterprises that intentionally seek to contribute solutions and measure progress to environmental and social challenges

Social Impact Bond (SIB) — A results-based contract in which one or more investors provide working capital for social programs, service providers (e.g., implementors) implement the program, and one or more outcome funders pay back the investors their principal plus a return if, and only if, these programs succeed in delivering results. In a social impact bond (SIB), the outcome payer is typically the government in high-income countries

Social Impact Incentive (SIINC) — This funding instrument rewards high-impact enterprises with premium payments for achieving social impact. The additional revenues enable enterprises to improve profitability and attract investment to scale

Social Success Note (SSN) — As a variation of an impact bond, also "pays for success" and crowds in commercial capital to finance social businesses. As with impact bonds, service providers receive capital from an investor upfront in the form of a loan, but instead of requiring the outcome funder (i.e., a donor government or foundation) to repay the investor for the loan provided, repayment is split between the outcome funder and the borrower. The borrower must pay back the principal of the loan with no interest and independent of the results obtained

Stakeholder — A party that has an interest in a company and can either affect or be affected by the business. The primary stakeholders in a typical corporation are its investors, employees, customers, and suppliers

Structured exits — A risk capital agreement where founders and funders contractually agree on a plan for the funder to fully (or partially) exit the investment. Unlike equity funders who have an open-ended agreement that relies on exponential growth and unknown future buyer or listing on a stock exchange, structured exit funders have a specific, achievable plan for how they are going to receive their return through dividends, profit sharing, redemptions, or a combination of repayment types

Sub-commercial capital — Investment capital or funding that expects to be invested and returned at below-market rates

Subordinated or junior debt — Subordinated or junior debt is most often an unsecured loan that ranks below other, more senior loans or securities with respect to claims on assets or earnings. In the case of borrower default, creditors who own subordinated debt will not be paid out until after senior bondholders/noteholders are paid in full. Subordinated debt may be secured by assets, but it is always second in line to senior debt holders in the case of default or underperformance of the borrower

Sustainable Development Goals (SDGs) — The 17 SDGs outlined by the United Nations are the blueprint to 2030 to achieve a better and more sustainable future for all. They address the global challenges we face, including poverty, inequality, climate change, environmental degradation, peace, and justice

Systems approach — An approach that brings together interviews, dialogue, openness to perspectives from public and private sectors, and people at all levels of an institution's hierarchy

Technical assistance (TA) — A type of blended finance where resources are used for skill building, capacity development, and/or consulting specific needs of a company or project

Theory of Change (ToC) — A schematic depicting the rationale and plan for achieving social and environmental outputs, outcomes, and impacts as a direct consequence of activities or services rendered

Total factor productivity (TFP) — A measure of productive efficiency in that it measures how much output can be produced from a certain amount of inputs

Transaction — An agreement, or communication, between a buyer and seller to exchange goods, services, or assets for payment

Venture capital — A form of private equity and type of financing that investors provide to startup companies and small businesses that are believed to have long-term growth potential

Selected Bibliography

Acumen Fund (2022) Investing as a Means: 20 Years of Patient Capital.

Argidius Foundation (2021) How to fulfill the potential of Business Development services using SCALE.

Argidius Foundation (2020) Digital Delivery: A Digitalization Guidebook for Enterprise Support Organizations. Aspen Network of Development Entrepreneurs.

Argidius Foundation (2019) Networking Works: Peer-to-Peer Business Networks Help Small and Growing Businesses Grow Revenues and Create Jobs.

BlueMark (2022) Making the Mark: Spotlighting Leadership in Impact Management.

BlueMark (2022) Raising the Bar: Aligning on the Key Elements of Impact Performance Reporting.

Clark C, Emerson J, Thornley B (2014) Collaborative Capitalism and the Rise of Impact Investing. Jossey-Bass, San Francisco.

Cohen, SR (2021) Impact: Reshaping Capitalism to Drive Real Change. Morgan James Publishing, New York.

Convergence (2021) The State of Blended Finance.

Dalberg (2022) Towards Market Transparency in Smallholder Finance: Early evidence from Sub-Saharan Africa.

Dalberg (2019) Bridging the credit gap for Micro and Small Enterprises through digitally enabled financing models.

Dalberg (2019) Closing the Gaps—Finance Pathways for Serving the Missing Middles. Collaborative for Frontier Finance.

Dalberg (2019) Unleashing Private Capital for Global Health. USAID.

Dalberg (2018) The Missing Middles: Segmenting Enterprises to Better Understand Their Financial Needs. Collaborative for Frontier Finance.

K. Hornberger, *Scaling Impact*,
https://doi.org/10.1007/978-3-031-22614-4

Dalberg (2018) Lessons from the Educate Girls Development Impact Bond.
Dalberg (2016) Mainstreaming Results-Based Finance: Actionable Recommendations for USAID.
Demirgüç-Kunt A, Klapper L, Singer D et al (2022) The Global Findex Database 2021: Financial Inclusion, Digital Payments, and Resilience in the Age of COVID-19. World Bank Group.
Dondi, M (2021) Outgrowing Capitalism—Rethinking Money to Reshape Society and Pursue Purpose. Fast Company Press, New York.
Endeavor Insight (2018) Fostering productive entrepreneurship communities: Key lessons on generating jobs, economic growth, and innovation.
GIIN (2021) Impact Investing Decision-making: Insights on Financial Performance.
Henderson, R (2020) Reimagining Capitalism in a World on Fire. PublicAffairs, New York.
IFC (2021) Using Blended Concessional Finance to Invest in Challenging Markets: Economic considerations, transparency, governance, and lessons of experience.
Impact Frontiers (2020) Impact-Financial Integration: A handbook for Investors.
MasterCard Foundation (2019) Pathway to Prosperity Rural and Agricultural Finance State of the Sector Report.
Mazzucato, M (2018) The Value of Everything: Making and Taking in the Global Economy. PublicAffairs, New York.
McKinsey (2020) Valuation: Measuring and Managing the Value of Companies. John Wiley & Sons, New York.
Morduch J, Beatriz A (2010) The Economics of Microfinance. MIT Press, Boston.
Nouwen C, Lee D, Hornberger K (2020) Five myths about impact bonds. India Development Review.
Novogratz, J (2020) Manifesto for a Moral Revolution: Practices to Build a Better World. Henry Holt, New York.
Pacific Community Ventures (2019) The Impact Due Diligence Guide.
Patton Power, A (2021) Adventure Finance—How to Create a Funding Journey That Blends Profit and Purpose. Palgrave MacMillan, London.
Pegon, Matthieu (2022) Mobilizing Beyond Leverage: Exploring the Catalytic Impact of Blended Finance, IDB Invest, Washington, DC.
Raworth, K (2017) Doughnut Economics: Seven Ways to Think Like a 21st-Century Economist. Chelsea Green Publishing, White River Junction.
Requarth, B (2020) VIVA the Entrepreneur: Founding, Scaling, and Raising Venture Capital in Latin America. Lioncrest Books, Austin.
Rottenberg, L (2014) Crazy Is a Compliment—The Power of Zigging When Everyone Else Zags. Penguin Group, New York.
Simon, M (2017) Real Impact: The New Economics of Social Change. PublicAffairs, New York.
World Bank (2018) A Guidebook for Effective Results-Based Financing Strategies.

Printed by Printforce, the Netherlands